Modularity and Twinning in Mineral Crystal Structures

Modularity and Twinning in Mineral Crystal Structures

Editor

Giovanni Ferraris

MDPI • Basel • Beijing • Wuhan • Barcelona • Belgrade • Manchester • Tokyo • Cluj • Tianjin

Editor
Giovanni Ferraris
Dipartimento di Scienze della Terra
Università di Torino
Turin
Italy

Editorial Office
MDPI
St. Alban-Anlage 66
4052 Basel, Switzerland

This is a reprint of articles from the Special Issue published online in the open access journal *Minerals* (ISSN 2075-163X) (available at: www.mdpi.com/journal/minerals/special_issues/MTMCS).

For citation purposes, cite each article independently as indicated on the article page online and as indicated below:

LastName, A.A.; LastName, B.B.; LastName, C.C. Article Title. *Journal Name* **Year**, *Volume Number*, Page Range.

ISBN 978-3-0365-1917-3 (Hbk)
ISBN 978-3-0365-1916-6 (PDF)

Contents

About the Editor . . . . **vii**

Preface to "Modularity and Twinning in Mineral Crystal Structures" . . . . **ix**

Giovanni Ferraris
Editorial for the Special Issue Modularity and Twinning in Mineral Crystal Structures
Reprinted from: *Minerals* **2021**, *11*, 758, doi:10.3390/min11070758 . . . . **1**

Ekhard K. H. Salje
Ferroelastic Twinning in Minerals: A Source of Trace Elements, Conductivity, and Unexpected
Piezoelectricity
Reprinted from: *Minerals* **2021**, *11*, 478, doi:10.3390/min11050478 . . . . **5**

Péter Németh
Diffraction Features from (101$\bar{1}$4) Calcite Twins Mimicking Crystallographic Ordering
Reprinted from: *Minerals* **2021**, *11*, 720, doi:10.3390/min11070720 . . . . **23**

Luca Bindi and Marta Morana
Twinning, Superstructure and Chemical Ordering in Spryite, $Ag_8(As^{3+}_{0.50}As^{5+}_{0.50})S_6$, at
Ultra-Low Temperature: An X-Ray Single-Crystal Study
Reprinted from: *Minerals* **2021**, *11*, 286, doi:10.3390/min11030286 . . . . **33**

**Sergey M. Aksenov, Alexey N. Kuznetsov, Andrey A. Antonov, Natalia A. Yamnova, Sergey
V. Krivovichev and Stefano Merlino**
Polytypism of Compounds with the General Formula $CsAl_2[TP_6O_{20}]$ (T = B, Al): OD
(Order-Disorder) Description, Topological Features, and DFT-Calculations
Reprinted from: *Minerals* **2021**, *11*, 708, doi:10.3390/min11070708 . . . . **43**

Emil Makovicky
Twinning of Tetrahedrite—OD Approach
Reprinted from: *Minerals* **2021**, *11*, 170, doi:10.3390/min11020170 . . . . **57**

**Anastasiia Topnikova, Elena Belokoneva, Olga Dimitrova, Anatoly Volkov and Dina
Deyneko**
$Rb_{1.66}Cs_{1.34}Tb[Si_{5.43}Ge_{0.57}O_{15}]\cdot H_2O$, a New Member of the OD-Family of Natural and Synthetic
Layered Silicates: Topology-Symmetry Analysis and Structure Prediction
Reprinted from: *Minerals* **2021**, *11*, 395, doi:10.3390/min11040395 . . . . **67**

Olga Yakubovich and Galina Kiriukhina
A Mero-Plesiotype Series of Vanadates, Arsenates, and Phosphates with Blocks Based on
Densely Packed Octahedral Layers as Repeating Modules
Reprinted from: *Minerals* **2021**, *11*, 273, doi:10.3390/min11030273 . . . . **81**

About the Editor

Giovanni Ferraris

Born in 1937 (Prarolo, Italy), Giovanni Ferraris obtained a master's degree in Physics at the University of Turin (1960) and a DSc (Libera docenza) in crystallography in 1969. He became a full professor of crystallography in 1975 at the Faculty of Sciences of Turin University. His main fields of research are hydrogen bonding, characterization of new mineral species, and modular crystallography. He is the (co-)author of 280 research papers published on peer reviewed journals, chapters of books, and of the book *Crystallography of Modular Materials* (Oxford University Press: 2004 and 2008) and has been the organizer of and speaker at national and international meetings and sections of general conferences. The high quality of his obtained results has been recognized by the following honors: Professor emeritus; dr. honoris causa at the universities of Bucharest (Romania) and Darmstadt (Germany); fellow of the National Academy of Lincei (Rome) and of the Academy of Sciences of Turin; foreign member of the Russian Academy of Natural Sciences.

Preface to "Modularity and Twinning in Mineral Crystal Structures"

In 2004, the book *Crystallography of Modular Materials* (IUCr/Oxford University Press) by G. Ferraris, E. Makovicky and S. Merlino (2nd edition 2008) systematically reviewed theory and examples of mineral crystal structures that can be described as built up by periodic repetition at the atomic scale of either one (planar) module (polytypes) or more modules—m1, m2, ...mn—with different compositions (polysomes). Whereas a series of polytypes, being based on the same module, essentially show a constant chemical composition, the members of a polysomatic series have different chemical compositions that depend on the m1/m2.../mn ratio. In a polysomatic series, the physical properties are a function of the chemistry of the modules and of their piling; thus, tailoring of the properties is possible. A special class of polytypes is rationalized by the so-called OD (Order/Disorder) theory. Twinning, i.e., the oriented association of two or more individuals of the same crystalline compound, is considered a modular structure at the macroscopic scale. The members of a series of polytypes or of polysomes usually have in common a supercell that, according to the reticular theory of twinning, favours the formation of twins.

Giovanni Ferraris
Editor

minerals

Editorial

Editorial for the Special Issue Modularity and Twinning in Mineral Crystal Structures

Giovanni Ferraris [1,2]

1   National Academy of Lincei, 00165 Rome, Italy; giovanni.ferraris@unito.it
2   Department of Earth Sciences, University of Turin, 10125 Turin, Italy

Citation: Ferraris, G. Editorial for the Special Issue Modularity and Twinning in Mineral Crystal Structures. *Minerals* **2021**, *11*, 758. https://doi.org/10.3390/min11070758

Received: 7 July 2021
Accepted: 13 July 2021
Published: 14 July 2021

Publisher's Note: MDPI stays neutral with regard to jurisdictional claims in published maps and institutional affiliations.

Ferraris et al. [1] systematically reviewed mineral phases whose crystal structures can be described as built up by periodic stacking at atomic scale of either one (usually) module (polytypes) or more modules—M, M', M" ... —with different compositions (polysomes). Whereas a series of polytypes shows essentially a constant chemical composition, the members of a polysomatic series have different chemical compositions that depend on the ratio M/M'/M" ... In a polysomatic series, the physical properties are a function of the chemistry of the modules and of their stacking; thus, tailoring of the properties is possible via synthesis of ad hoc compounds. A special class of polytypes is rationalized by the so-called order/disorder (OD) theory [2] that plays a basic role in most of the articles published in this special issue of Minerals.

Twinning, i.e., the oriented association of two or more crystals of a same compound, is considered a modular structure at macroscopic scale (see Chapter 5 in [1]). The members of a series both of polytypes and of polysomes usually have in common a supercell that, according to the reticular theory of twinning, favors the formation of twins.

The articles published in this issue show that polytypism, polysomatism, and twinning are fertile fields of research and their basic principles are powerful tools to solve and classify related crystal structures. The detection of twinned crystals is almost no longer a matter of morphological observation as it used to be in the past. This task is now left to algorithms that are routinely included in the software dedicated to the solution and refinement of the crystal structures.

However, research on twinning and its consequences on structure and properties of crystalline materials is a growing field and a frontier field concerning twin walls is reviewed in this issue by Salje [3]. The author shows that a specific type of twinning is a very common phenomenon in ferroelastic materials where the thin areas (twin walls or twin boundaries) between the twin domains yield characteristic physical and chemical properties. This category of crystal structures generates a large variety of properties. Some properties of the twin walls are discussed in the review article, such as their ability for chemical storage, and their structural deformations which generate polarity and piezoelectricity inside the walls. It is noteworthy that none of these effects exist in the adjacent domains of the twins because only their twin walls contain topological defects, such as kinks, and are strong enough to deform surface regions.

Instead, classical twinning is central in the articles contributed by Németh [4], Bindi and Morana [5], and Makovicky [6]. Via the analysis of electron diffraction patterns and high-resolution TEM images, the article [4] shows that the extra reflections occurring halfway between the Bragg reflections of glendonite, a calcite ($CaCO3$) pseudomorph after ikaite ($CaCO_3 \cdot 6H_2O$), are the consequence of microtwinning by reticular pseudomerohedry based on a pseudo-orthorhombic C-centered sublattice. The same diffraction features correspond to the so-called carbonate c-type reflections associated with Mg and Ca ordering, a phenomenon that cannot occur in pure calcite, as is the studied sample.

The article by Bindi and Morana [5] deals with the structural study of a low-temperature phase of the mineral spryite, $(Ag_{7.98}Cu_{0.05})_{\Sigma\,=\,8.03}(As^{5+}_{0.31}Ge_{0.36}As^{3+}_{0.31}Fe^{3+}_{0.02})_{\Sigma\,=\,1.00}S_{5.97}$, a new natural sulfosalt recently described. At room temperature, spryite is orthorhombic space group $Pna2_1$ with cell parameters $a = 14.984$, $b = 7.474$, $c = 10.571$ Å; in its structure As^{3+} and As^{5+} statistically occupy the same crystallographic position. X-ray diffraction data were collected at 30 K (helium cryostat) on the same sample previously characterized from a chemical and structural point of view at room temperature. At 30 K, spryite maintains the $Pna2_1$ space group, but shows an $a \times 3b \times c$ superstructure and twinning by reticular merohedry based on a hexagonal sublattice. The superstructure arises from the ordering of As^{3+} and (As^{5+}, Ge^{4+}) in different crystal–chemical environments. Recognition of twinning allowed solving and refining the crystal structure.

The twinning by merohedry of tetrahedrite–tennantite (point group $-43m$) is well known. In his article, Makovicky [6] investigates the structural aspect of this twinning and finds that it is related to defects which can occur during the growth of tetrahedrite crystals; consequently, the twinning is described as an order–disorder (OD) phenomenon. The author takes as unit OD layer a one-tetrahedron-thick (111) layer composed of six-member rings of tetrahedra that are stacked according to the sequence ABCABC. The sequence can be described by three consecutive tetrahedron configurations, named α, β, and γ. When the orientation of the component tetrahedra is uniform, the α, β, γ, α sequence builds the cage structure of tetrahedrite. If it happens that the tetrahedra of the β layer are rotated by 180° against those in the underlying α configurations and/or rotated α configuration follows after the β configuration (instead of γ), the defect allows the growth of the crystal structure according to a different orientation, i.e., a twinning is generated.

The use of the OD theory is also central in the articles by Aksenov et al. [7] and Topnikova et al. [8]. In [7], the crystal structures of compounds with the general formula $Cs\{^{[6]}Al_2[^{[4]}TP_6O_{20}]\}$ (T = Al, B) are analyzed. Microporous, heteropolyhedral MT frameworks, consisting of octahedra (M) and tetrahedra (T), occur in these structures; thus, they are suitable for the migration of small cations such as Li^+, Na^+, and Ag^+. The studied structures display order–disorder (OD) character and are described using the same OD groupoid family and two kinds of nonpolar layers that alternate along the **b** direction and have common translation vectors **a** and **c**. Both ordered polytypes and disordered structures are obtained using the following symmetry operators: a 2_1 screw axis parallel to **c**, inversion centers, and a screw axis parallel to **a**.

Article [8] reports the crystal structure of a new silicate–germanate with formula $Rb_{1.66}Cs_{1.34}Tb[Si_{5.43}Ge_{0.57}O_{15}] \cdot H_2O$. However, for this special issue, the important part of the article is represented by the comparison of the investigated structure with those of other related layered natural and synthetic silicates. The comparison is based on a topology–symmetry analysis by OD approach. A wollastonite chain was selected as initial structural unit. Three symmetrical ways of forming ribbons from such chain and three ways of interconnecting the ribbons to form layers are described using symmetry groupoids. Finally, hypothetical variants of the described layers and ribbons are predicted.

In the last article [9], a family of layered vanadates, arsenates, and phosphates, including the minerals vésignéite and bayldonite, and synthetic analogous is investigated. Their structures are built by similar modules that consist of a central octahedral layer filled by Mn, Ni, Cu, or Co and adjacent $[VO_4]$, $[AsO_4]$, or $[PO_4]$ tetrahedra. These structures are discussed in terms of modular crystallography and are shown to form a mero-plesiotype polysomatic series, i.e., a series where a module is common to all the members of the series, whereas a second module is peculiar for each member (merotypy) and, at the same time, the common layer may be modified both in its chemical composition and, to some extent, in its topology (plesiotypy; see Chapters 1 and 4 of [1]).

The articles published in this issue offer a sound update of research fields where the concepts of modular crystallography are well alive and provide unique tools to understand and to solve problems of structural crystallography. To note that even if the OD theory is quite often applied to mineral phases, it has been successfully applied to molecular crystals too, as, e.g., exemplified by the articles [10–12]. Reference [11], besides reporting the OD character of the crystals of 9,10 phenantrenequinone, illustrates the basic elements of the OD theory.

Data Availability Statement: Not applicable.

Conflicts of Interest: The author declares no conflict of interest.

References

1. Ferraris, G.; Makovicky, E.; Merlino, S. *Crystallography of Modular Materials*; Oxford University Press: Oxford, UK, 2004; ISBN 0-19-852664-4.
2. Dornberger-Schiff, K.; Grell-Niemann, H. On the theory of order–disorder (OD) structures. *Acta Crystallogr.* **1961**, *14*, 167–177. [CrossRef]
3. Salje, E.K.H. Ferroelastic Twinning in Minerals: A source of trace elements, conductivity, and unexpected piezoelectricity. *Minerals* **2021**, *11*, 478. [CrossRef]
4. Németh, P. Diffraction Features from (10$\bar{1}$4) Calcite Twins Mimicking Crystallographic Ordering. *Minerals* **2021**, *11*, 720. [CrossRef]
5. Bindi, L.; Morana, M. Twinning, Superstructure and Chemical Ordering in Spryite, $Ag_8(As^{3+}_{0.50}As^{5+}_{0.50})S_6$, at ultra-low Temperature: An X-ray Single-Crystal Study. *Minerals* **2021**, *11*, 286. [CrossRef]
6. Makovicky, E. Twinning of Tetrahedrite—OD Approach. *Minerals* **2021**, *11*, 170. [CrossRef]
7. Aksenov, S.M.; Kuznetsov, A.N.; Antonov, A.A.; Yamnova, N.A.; Krivovichev, S.V.; Merlino, S. Polytypism of compounds with the general formula $Cs\{Al_2[TP_6O_{20}]\}$ (T = B, Al): OD (order-disorder) description, topological features, and DFT-calculations. *Minerals* **2021**, *11*, 708. [CrossRef]
8. Topnikova, A.; Belokoneva, E.; Dimitrova, O.; Volkov, A.; Deyneko, D. $Rb_{1.66}Cs_{1.34}Tb[Si_{5.43}Ge_{0.57}O_{15}]\cdot H_2O$, a new member of the OD-family of natural and synthetic layered silicates: Topology—symmetry analysis and structure prediction. *Minerals* **2021**, *11*, 395. [CrossRef]
9. Yakubovich, O.; Kiriukhina, G. A Mero-plesiotype Series of Vanadates, Arsenates, and Phosphates with Blocks Based on Densely Packed Octahedral Layers as Repeating Modules. *Minerals* **2021**, *11*, 273. [CrossRef]
10. Kautny, P.; Schwartz, T.; Stöger, B.; Fröhlich, J. An unusual case of OD-allotwinning: 9,9′-(2,5-dibromo-1,4-phenylene)bis[9H-carbazole]. *Acta Crystallogr.* **2017**, *B73*, 65–73. [CrossRef]
11. Merlino, S. OD approach to polytypism: Examples, problems, indications. *Z. Krist.* **2009**, *224*, 251–260. [CrossRef]
12. Merlino, S. OD character and polytypic features of the structure of the molecular crystal: (1R,3S)-dimethyl 2-oxocyclohexane-1,3-dicarboxylate. *Atti Soc. Toscana Sci. Nat. Mem. Ser. A* **2016**, *123*, 61–65.

Review

Ferroelastic Twinning in Minerals: A Source of Trace Elements, Conductivity, and Unexpected Piezoelectricity

Ekhard K. H. Salje

Department of Earth Sciences, University of Cambridge, Cambridge CB2 3EQ, UK; ekhard@esc.cam.ac.uk

Abstract: Ferroelastic twinning in minerals is a very common phenomenon. The twin laws follow simple symmetry rules and they are observed in minerals, like feldspar, palmierite, leucite, perovskite, and so forth. The major discovery over the last two decades was that the thin areas between the twins yield characteristic physical and chemical properties, but not the twins themselves. Research greatly focusses on these twin walls (or 'twin boundaries'); therefore, because they possess different crystal structures and generate a large variety of 'emerging' properties. Research on wall properties has largely overshadowed research on twin domains. Some wall properties are discussed in this short review, such as their ability for chemical storage, and their structural deformations that generate polarity and piezoelectricity inside the walls, while none of these effects exist in the adjacent domains. Walls contain topological defects, like kinks, and they are strong enough to deform surface regions. These effects have triggered major research initiatives that go well beyond the realm of mineralogy and crystallography. Future work is expected to discover other twin configurations, such as co-elastic twins in quartz and growth twins in other minerals.

Keywords: twin wall; twin boundary; minerals; emerging properties; piezoelectricity in minerals; surface relaxations; anorthite; Pamierite; perovskite

 check for updates

Citation: Salje, E.K.H. Ferroelastic Twinning in Minerals: A Source of Trace Elements, Conductivity, and Unexpected Piezoelectricity. *Minerals* **2021**, *11*, 478. https://doi.org/10.3390/min11050478

Academic Editor: Giovanni Ferraris

Received: 30 March 2021
Accepted: 22 April 2021
Published: 30 April 2021

Publisher's Note: MDPI stays neutral with regard to jurisdictional claims in published maps and institutional affiliations.

1. Introduction

Twinning of minerals has intrigued mineralogists as long as mineralogy was established as a discipline by Haüy in 1797 [1]. At the beginning of the 20th century, many textbooks and research articles in mineralogy focused on twinning from a geometrical perspective [2–8]. Many of the early pioneers of mineralogy and crystallography were involved in this research, which highlights its importance. Landmark contributions came from Hugo Strunz [9] and Paul Ramdohr [10], who were inspirational for generations of mineralogists. The impact of work on the classification of 'twins' declined over recent decades; however, while research on the consequences of twinning sharply increased in a wider field outside mineralogy, namely physics, chemistry, ferroic materials, and nanotechnology. More than 8000 papers on twinning were published in 2020 alone. The fundamental shift came with a change of focus from twins to twin walls. A simple experiment partially initiated the new perspective, which was first disbelieved but then confirmed and extended to many other scenarios. The experimental discovery was that, in a specific tungstate mineral with perovskite structure, WO_3, the boundary between various twins, the so-called 'twin boundary' or 'twin wall', has completely different physical properties from the twin domains (Figure 1). In this particular case, the twins are insulators while the twin boundaries are highly conducting at room temperature and become superconducting below ca. 3 K [11].

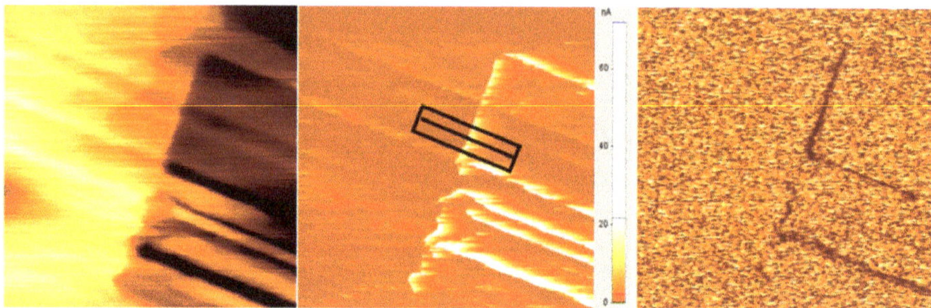

Figure 1. Topology (left), conductivity (middle), and piezoelectricity (right) of twin structures in WO_3 at room temperature. The structure is tetragonal P-42$_1$m and piezoelectric [12]. Twin walls appear as borders between twin domains on the right, and as bright conducting regions in the middle panel. The piezoelectricity is short-circuited in the highly conducting domain walls, which now appear dark in the PFM image on the right (horizontal length of the image 1.5 µm, after [13]).

This review will describe other properties, and they were later identified to be localized in the twin walls, but not in the twin domains. These properties are highly relevant for technological applications and they triggered the shift from research on twin domains to research on domain walls. Hence, the fundamental new discovery is that twin walls are no simply thin layers inside minerals with essentially the same properties as the twin domains. Instead, they represent novel materials, confined to the twin walls, which do not exist as domains. Therefore, it is wrong to presume that the atomic configurations in the twin walls are the same as in the twins. This realization is similar to the earlier observation that surfaces are special because they contain surface reconstructions and surface relaxations, which often lead to the discovery of important new structures that are fundamentally different from the bulk. The earlier notion that twin walls are like internal surfaces, rather than infinitely thin layers of glue between twin domains, hence has found some late justification. The difference between twin boundaries and surfaces is that minerals can contain a large number of twin boundaries, while surface regions are necessarily limited under most circumstances. Hence, the emphasis of current research is to control twinning and use the twin walls as active elements. Applications are superconducting layers, transistors, photoelectric detectors, memory devices, piezoelectric layers, sources of chemical elements for reactions in confined spaces, and many more. The key notion is that wall properties are fundamentally different from the bulk properties. These physical, chemical, and structural properties are typically 'emerging', i.e., they exclusively relate to the features of the twin wall. They disappear when the twin walls disappear, so that mono-domain samples do not display any of these emerging properties.

The move towards research on specific micro- and nano-structures that largely started in 2010 with a new research discipline of 'domain boundary engineering' [14,15]. The aim was to understand how to generate twinned materials with desirable twin wall effects. A recent review by Nataf et al. [16,17] summarizes many aspects of successes and failures in 'domain wall engineering' from a physics perspective. From a mineralogical perspective many of these effects were initially discovered in minerals with naturally occurring twin walls. We will now review some examples of minerals where important wall properties were found. We will then argue that much more work needs to be done on different twins in other minerals and highlight some open questions.

2. Ferroelastic Twin Laws in Minerals

Typical examples for phase transitions in minerals, which constitute large proportions of the earth's crust, are those of quartz and feldspar. Quartz transforms from a high-temperature form with *622* symmetry (β-phase) via one or two incommensurate phases (INC) into the trigonal a phase with *32* symmetry (α-phase) [18]. The transition mechanism is related to the rotation of SiO_4 tetrahedra with additional small distortions of the tetrahedral bond lengths and bond angles. The symmetry change during the phase transition is related to the appearance of a piezoelectric symmetry tensor, so that the transition cannot be ferroelastic. No ferroelastic twin laws are symmetry allowed to explain the multiple twin structures, because quartz is, instead, co-elastic. The inner structure of its twin walls in quartz has not yet been fully explored, and it still constitutes a major challenge to mineralogists [19,20]. For example, first principle DFT calculations are highly desirable for the determination of local crystal structures in twin walls of quartz. Co-elastic minerals are defined as undergoing strain deformations during phase transitions that are not confined to the symmetry breaking required for ferroelasticity. The strain often is a volume strain, like in the case of quartz, which does not lead to twinning, but allows twin-like structures inside the incommensurate phases. Nevertheless, co-elastic minerals are often twinned for other reasons than ferroelasticity, such as growth twinning, twinning under uniaxial stress, etc. Their twin boundaries are still awaiting a major research effort to understand their local structures.

This behavior contrasts with the structurally more intricate transitions in feldspars. Their transitions are related to three mechanisms: (i) the ordering of Al and Si in a tetrahedral network, (ii) structural distortions of the network of an essentially displacive nature (with a variety of critical points in the Brillouin zone), and (iii) the ordering/exsolution of the alkali and earth-alkali atoms. Na-feldspar is monoclinic $C2/c$ at high temperatures and triclinic P-1 at low temperatures. The phase transition is related to the simultaneous symmetry breaking of the monoclinic structure via a displacive lattice distortion and a simultaneous ordering of the Al and Si atoms. At lower temperatures (around 950 K), the contributions of these two processes change in a crossover mechanism [21,22]. Two order parameters characterize the transition behavior. A weak thermodynamic singularity occurs due to a second-order transition with predominantly displacive character due to the tilt of the structural crankshafts. This lattice distortion drags with it a cation ordering process, which itself drives a crossover mechanism that operates without actual symmetry breaking at lower temperatures [23]. The coupling between several structural instabilities is very common in mineral systems and may, at first sight, seem to be rather complicated. Nevertheless, both of the mechanisms follow the same ferroelastic symmetry breaking process from the monoclinic to the triclinic phase. This means that the twin laws can be derived uniquely from symmetry conditions. The orientation of ferroelastic twins is then fully determined by the deformation of the P-1 phase relative to the $C2/c$ phase. A full analysis of albite and pericline twin laws in alkali feldspars in [24] demonstrates the power of this method.

A complete description of all possible ferroelastic transition symmetries with many examples of minerals can be found in a textbook [25] and a specific review for the application of Landau theory in minerals [26]. A classification of symmetry conditions for the strain that determines twin laws was published by Carpenter and collaborators [27–29]. They derived a formal framework for detailed observations of the dynamical behavior of twin structures.

The orientation of ferroelastic twins is determined by the Sapriel conditions that twin walls are strain free [30]. Only two possible types of walls between twins are symmetry allowed. They are either completely fixed by the crystallography of the low symmetry phase or they are not. Fixed walls are called w-walls and flexible walls are called w'-walls. W-walls are related to mirror planes that are lost during the phase transition, while w'-walls contain equivalent diads inside the walls. The mineral palmierite is a typical example. Figure 2 shows the twin structure of a palmierite sample with composition $Pb_3(AsO_4)_2$, which underwent a phase transition R-$3m$ to $C2/m$. During the transition, the sample developed twins with w and w' twin walls. The image shown in Figure 2 was taken along the pseudo-triad, which explains the almost threefold symmetry of the twin walls. The exact angles between the twin walls follow directly from the Sapriel condition and they deviate from the multiples of 90° and 30°. In 2020, Yokota et al. [31] found that all of the twin walls in palmierite are piezoelectric, while the twin domains are not. The structural symmetry inside the twin walls, namely m and 2, was confirmed.

W walls in Figure 2 appear as sharp lines because they are along the lost mirror plane of the *R-3m* phase which are perpendicular to the surface. W' walls are inclined to the surface and change their orientation with temperature. They contain the broken diads of the high symmetry phase. The angles between the twin walls deviate from the multiples of 90° and 30° because the spontaneous strain deforms the twin orientations. All orientations were reproduced from the Sapriel conditions that twin walls are unstrained in [32].

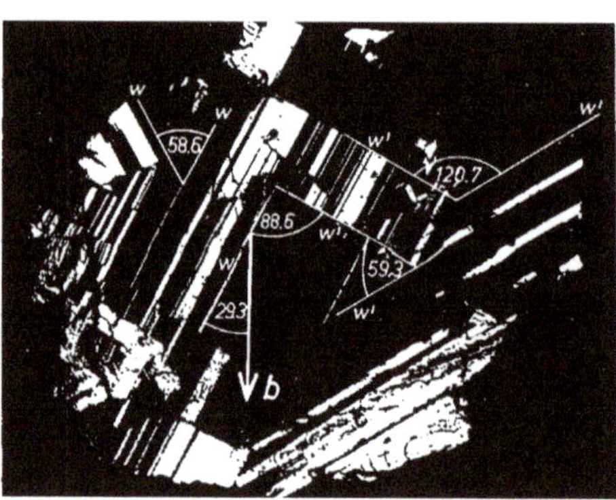

Figure 2. Birefringence image of the twin structure of the palmierite $Pb_3(AsO_4)_2$ in the *C2/m* phase. Twin variants appear as black and white regions. The birefringence image was taken along the pseudo-trigonal axis with some twin domains in extinction direction (after [32]).

3. Twin Walls as Storages for Cations and Their Pinning Behaviour in Anorthoclase

Anorthoclase is a prototypic mineral for local exchanges of cations in twin walls. Inscribing and erasing twin domains (so-called 'twin memory' and 'twin amnesia') was experimentally observed in anorthoclase with composition Ab70 Or25 An5 from volcanic tuffs in Camperdown, Victoria (Sample 195127 in the Harker Collection, Cambridge University) [33]. After collecting the room temperature XRD rocking curve of anorthoclase, a sample was heated rapidly in vacuum to some high temperature above the displacive transition temperature for the sample (733 K); this temperature was then maintained for 4 h. The sample was then cooled back to room temperature (with a cooling rate of ~10 K·min^{-1}), where the rocking curve was again recorded. The cycle of collecting a room temperature rocking curve and heating to progressively higher and higher temperatures was repeated until an annealing temperature of ~1000 K was reached.

Rocking curves clearly reveal the formation of twins. They showed that twins were generated by quenching the sample through the phase transition temperature. This leads to the 'memory effect': the twin structure remains stable and it will always reappear at exactly the same locations in the sample when reheated and re-cooled through the transitions. This effect continues until the walls are erased or overwritten by further heating, which qualifies as 'twin amnesia'. Annealing at 860 K does not induce much twin amnesia, while annealing at 880 K has a much greater effect. The phase diagram for anorthoclase reveals the underlying mechanism. The temperature at which twin memory begins to be lost is near the solvus temperature for anorthoclase with the composition of the Camperdown sample, where segregation between Na and K commences on cooling. Thus, the diffusion K into the twin boundaries was identified as mechanism for the twin memory effect, while the replacement by Na induces amnesia (Figures 3 and 4).

The local crystal structure inside the twin boundary differs from that in the bulk [34–38], and this has the effect of attracting Na or K to the twin boundary. Simple symmetry and volume arguments favor the larger K ions on the twin boundaries, and Na in the bulk.

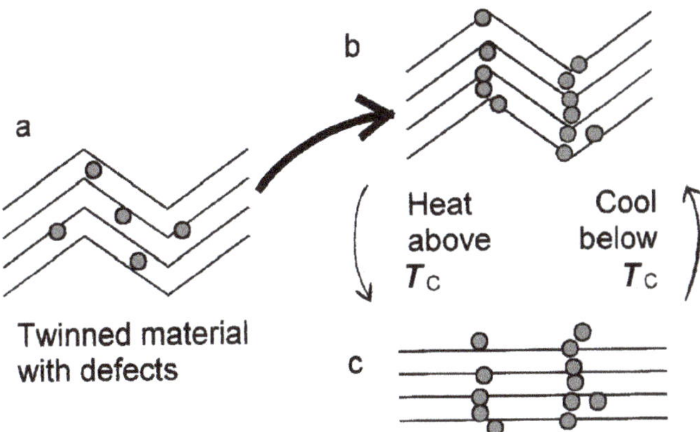

Figure 3. Schematic illustration of atomic-scale mechanism of twin memory in anorthoclase (after [34]).

While alkali segregation is the mechanism that is responsible for twin memory in anorthoclase, one expects memory loss to occur quite rapidly once the solvus temperature is exceeded, since alkali diffusion through the feldspar structure is rapid [39], and the diffusion lengths that are required for twin memory are quite small. At room temperature, the twin wall thickness in anorthoclase is of the order of 2.5 nm [40]. Thus, there are sufficient potassium cations to completely fill the twin wall in a relatively narrow layer with a maximum diffusion length required to be of the wall thickness.

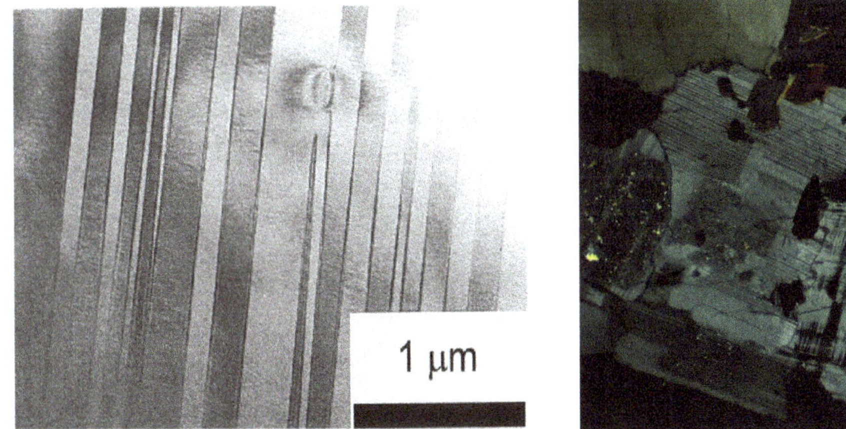

Figure 4. Typical TEM micrographs of anorthoclase. This image at the left is of a (010) slice of Camperdown anorthoclase, and shows the pericline twins approximately parallel to (001). The image at the right shows a wider view of twinning in an albite rich plagioclase in granite. Twins are seen as narrow lines with two dominant orientations, representing the albite and pericline twin laws. The diameter of the field of view is 3 mm (courtesy Prof. M.A. Carpenter).

This example highlights the effect of 'loading' and 'unloading' twin walls with cations or defects. In the case of anorthoclase, the change of chemical composition inside twin walls was already observed in 2000 by microprobe techniques [36]. Such chemical changes, in turn, pin the position of twin walls, so that the twin walls appear to be static under weak fields while the un-doped sample shows freely moving twin boundaries. This effect was also simulated by computational molecular dynamics methods [41]. The results show that randomly distributed, static defects are enriched in ferroelastic domain walls. Quantitatively, the relative concentration of defects in walls, Nd, follows a power law distribution as a function of the total defect concentration C: $N_d \sim C^\alpha$ with $\alpha = 0.4$. In these simulations the enrichment Nd/C ranges from ca. 50 times to ~3 times. The enrichment is due to the nucleation of twins at defect sites; the dynamics of twin nucleation and switching are then strongly dependent on the defect concentration. Under stress, the domain walls do no longer move smoothly, but form avalanches. This phenomenon is rather common when walls de-pin and nucleate new walls, see the reviews in [16,42–46]. The energy distributions of these avalanches follow power laws with energy exponents ε between 1.33 and 2.7. These exponents then allow for a detailed characterization of the pinning behavior and assessment how much the twin structures adapt to the defect distribution (Figures 5 and 6).

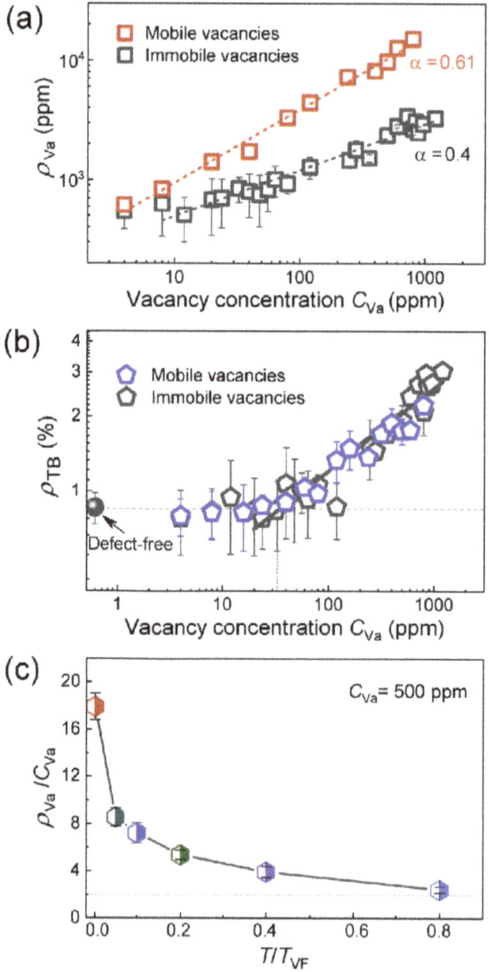

Figure 5. (**a**) The relative concentration of vacancies at boundaries r_{Va} as a function of the vacancy concentration C_{Va} at $T = 5 \times 10^{-4}$ T_{VF}. T_{VF} is the Vogel-Fulcher temperature of the twin mobility. For mobile vacancies, after long time diffusion, it follows a power law $r_{Va} \sim C_{Va}^{\alpha}$ with $\alpha = 0.61$ (red data). The grey square shows the immobile vacancy. (**b**) The variation of twin boundary density r_{TB} with C_{Va} at $T = 5 \times 10^{-4}$ T_{VF} follows a power law relationship. As $r_{TB} \sim C_{Va}^{\alpha}$ with $\alpha = 1/3$. (**c**) The enrichment ratio of r_{Va}/C_{Va} decreases with increasing temperature and drops to ca. 2.45 at $T = 0.8T_{VF}$. Reproduced with permission from [47] published by Elsevier Ltd., 2019.

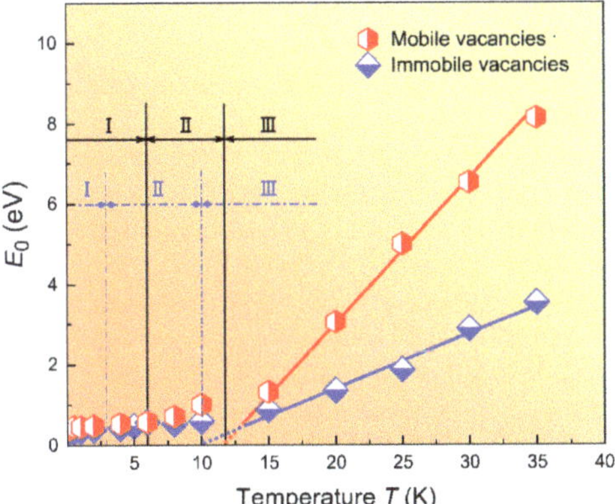

Figure 6. Phase diagram for the energy E_0 of mobile and immobile vacancies as a function of temperature. E_0 is defined as $E_0 \sim k_B(T - T_{VF})$, which corresponds to an activation energy at the temperature T. The Vogel–Fulcher temperature is T_{VF}, which is near 12K in this model. Reproduced with permission from [47] published by Elsevier Ltd., 2019.

This mechanism changes when defects have mobilities that are comparable with those of mobile twin walls [47]. Molecular dynamics analysis indicates how vacancies reduce their energy by residing in twin boundaries, kinks inside twin walls, and junctions between twin walls. Vacancies have the largest binding energy inside junctions and they co-migrate with the motion of the junctions between twin walls. For defects trapped inside twin walls, a "ghost line" may be generated, because vacancies do not necessarily diffuse with moving boundaries, similar to the memory effect in anorthoclase. They leave a trace of their previous position when they are left behind. Needle twin walls act as channels for fast diffusion with almost one order of magnitude higher vacancy diffusivity than in the bulk. This modifies the relative concentration of vacancies at twin boundaries as a function of the average vacancy concentration. The concentration of vacancies at twin boundaries is enriched ca. five times at low temperatures, as determined by these simple simulations. With increasing temperature, the enrichment drops as the trapping potential at the twin boundaries decreases via thermal release. The distribution of energy-drops upon twin pattern evolution follows a power law. The energy exponent of the mobile twin walls ε increases from ~1.44 to 2.0 when the vacancy concentration increases. Such enrichments of cations and vacancies in twin walls have been widely reported in the literature (e.g., in perovskites [37], klinker [48], twinned alloys [49], pure metal twin walls [50], cassiterite (SnO_2) [51], tellurized molydenite [52], tungsten dichalcogenites [53], and twin wall doping in palmierite [54]. Oxygen vacancy trapping in perovskite was simulated and the results are discussed in [55,56]. The determination of chemical variabilities of twin boundaries have become an active field of research, both experimentally e.g., when trace elements are hidden in twin walls, and by computer simulation.

The chemical changes in twin walls requires that the twin walls have a finite volume. It can be estimated that the typical proportion of atoms in twin walls in a heavily twinned mineral is ca 1–10 ppm. The thickness of the twin walls has been measured for several minerals and it extends over some nanometers (2 nm in palmierite [57], 1.3 nm in anorthoclase [40], 1.6 nm in tungstate [58], 1.6 nm in $LaAlO_3$, [59], and 2 nm in perovskite [60]). In addition, the possibility of thickness measurements by surface methods was raised [61–63]. It was argued that twin wall thicknesses will vary significantly with their local chemical composition [64,65]. Nevertheless, the length scale over which the twin walls expand or retract, their thickness remains below 10 nm. Hence, experimental work to quantify this variation is very hard to perform, because other effects, like kink-formation and wiggles in twin walls, will be superimposed onto the compositional effect and may obscure them.

4. Structural Changes and Electric Polarization inside Twin Boundaries in the Mineral Perovskite

Even when twin walls are chemically inert and do not change their composition relative to the twin domains, they will still modify their crystal structure. The simple reason is that the unit cells in twin walls are systematically deformed by the strain that is exerted by the twin domains. A simple sketch presented in Figure 7 shows a simplified lattice configuration with two twins. The repetition lengths between the lattice points have the same length as in the twin domain and in the wall (tainted red in Figure 7). Only the second nearest neighbor lengths across the twin wall are different in the wall from the bulk. This leads to subtle structural differences in the domain and walls. A large cation is then imagined to be inserted in these local cells. While their optimal position in the rectangles is near its midpoint, this is not true for the wall. Here, the atoms will shift towards the upper end of the kite-shapes cells. This simple thought experiment illustrates why large atoms in the twin walls are always shifted with respect to the twin domains, and that these shifts commonly are directed towards the apex of the wall cells.

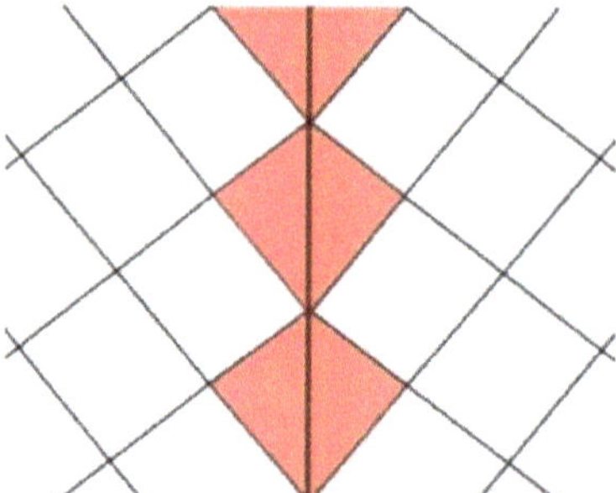

Figure 7. A simple model of a twin wall where the exchange between short and long repetition lengths between second nearest neighbor lattice points leads to a lattice distortion inside the red wall area.

This twin wall effect has been predicted in much more detail using Landau theory [66–68]. The structural deviations in the twin walls are substantial and they play a major role in device fabrication [69–71] and so-called adaptive materials where the phase equilibria are achieved by suitable twinning processes [72]. A typical example is the twin structure in the mineral perovskite ($CaTiO_3$), where the intrinsic wall activation energy was first calculated by Lee et al. [73]. The breakthrough for the detailed structure determination was achieved by Van Aert et al. [60], who demonstrated, by transmission electron microscopy, that the Ti positions in the twin walls were indeed shifted by 6 pm along the wall direction (Figures 8 and 9). Such large shifts were previously unexpected, and they rank amongst the biggest polar shifts in any mineral. It is comparable with the ferroelectric shift of Ti in the ferroelectric material $BaTiO_3$.

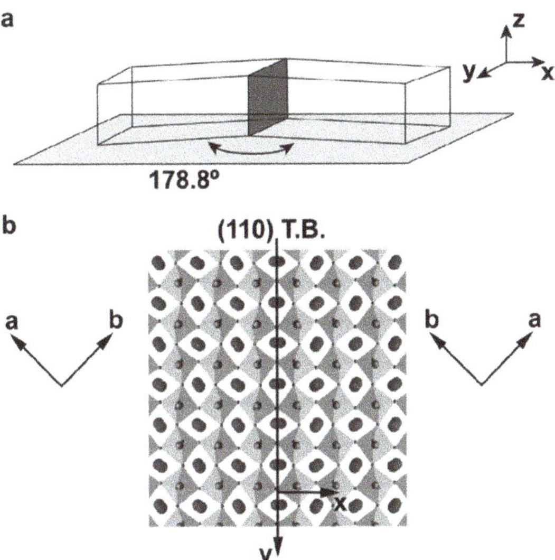

Figure 8. Perovskite CaTiO$_3$ (**a**) Single twin wall, indicated as standing dark grey plane, with the chosen (x, y, z) reference system for the definitions of the measured displacements. The angle of 178.8° is the result of the twinning operation from the high temperature cubic to the low temperature orthorhombic phase. (**b**) Atomic configuration on both sides of the (110) twin plane with Ca atoms marked by large, filled circles, Ti atoms by medium-sized shaded circles, and O atoms by small circles. Reproduced with permission from [60], published by John Wiley & Sons, Inc., 2011.

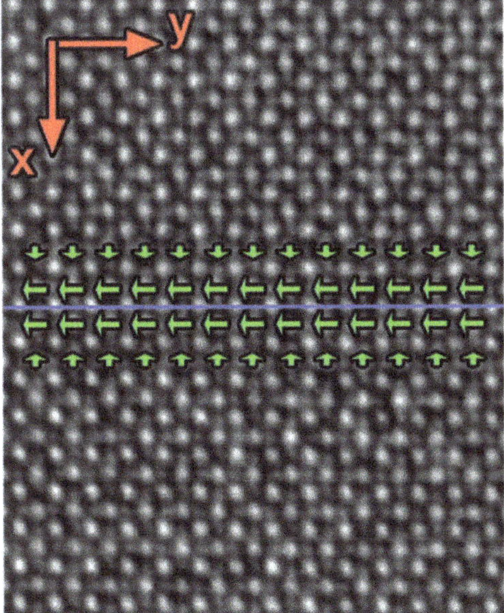

Figure 9. Mean displacements of the Ti atomic columns from the centre of the four neighbouring Ca atomic columns inside the twin wall are indicated by green arrows in perovskite CaTiO$_3$. The centre row of atoms shifts to the left and generates polarity of the domain wall. The polarization perpendicular to the wall is self-compensating. The blue line indicates the middle of the twin boundary. Reproduced with permission from [60], published by John Wiley & Sons, Inc., 2011.

The dramatic structural deformation inside twin walls has been observed in many materials, although it was first discovered in the mineral perovskite. Technically, the measurement of symmetry breaking in twin walls is very involved using transmission electron microscopy. It is greatly eased by using the second harmonic generation, SHG technique, i.e., the observation of the doubling of the laser frequency in piezoelectric twin walls, as a measure for the local structural distortion. This technique is ideal if the twin domains are centrosymmetric, while the symmetry breaking renders the twin walls piezoelectric. SHG does not see the twin domains in this case, but it gives strong signals on the twin walls [74]. The symmetry relations of the SHG allow for the determination of local point symmetry of the twin wall. Another useful method is phonon spectroscopy [75,76]. Polar twin walls were also confirmed in other minerals, such as palmierite [31,54] and clinobisvanite BiVO$_4$ [77,78]. Very high densities of twin walls, such as in tweed structures lead to an overall SHG background that proves that the mineral contains an extremely high density of walls that is hard to see using other techniques [79]. The tweed structure, as an extension of the twin structure, was already predicted in the 1990s [80–85] and recently observed in LaAlO$_3$. A short review on tweed in minerals can be found in [86]. A typical selection of tweed in minerals is given in [87–92]. However, none of these minerals have yet been analyzed in sufficient detail to quantify the local symmetry breaking effects in the twin walls or in the tweed structure.

5. Topological Changes of Twin Walls: Kinks and Surface Intersections

Twin walls are planar in local areas, but not globally. They are prone to topological defects, like kinks, arching, and systematic changes, when they touch surfaces. In addition, they deform dynamically under external stress. This effect is at the core of Resonant Ultrasonic Spectroscopy, RUS. Its application to twins in minerals is far too extensive to be reviewed; here, the reader is referred to the excellent review by Carpenter and Zhang [93] which covers several minerals. Specific examples were discussed [94–99]. The underlying mechanisms of these dynamic phenomena of twins were computer simulated [100–102] and they show a multitude of mechanisms like the progression and retraction of needles, wall arching and shifts of wall defects [100–103]. They analyzed the movement of needles, domains and wall bending of twin domains, which is retarded by interactions with the crystals structure. These interactions lead to pinning and, hence, to the damping of the wall movement. RUS experiments prove that twin walls are locally mobile, and that damping can be measured quantitatively. Hence, the macroscopic elastic response of a twinned ferroelastic mineral has very little to do with its intrinsic elastic properties, but is greatly reduced by the spatial relaxations and energy absorption of twin walls. Only when the uniaxial stress is large enough to eliminate all twins, will one be able to measure the intrinsic, un-twinned elastic properties. As the scientific question has now turned towards the understanding of twins and their boundaries, elastic measurements need to avoid these excessive stresses. Typical external strains in elastic measurements are in the order of 10^{-7} to 10^{-4}, i.e., being much lower than typical detwinning strains [104]. The induced wall movements are equally small with resonance amplitudes in the order of several attometers. Only the walls between twin domains contribute to damping, but not the twin domains themselves, which gives excellent access to wall data that are not polluted by bulk effects.

Common perturbations of twin walls are kinks or latches. Figure 10 shows a typical example.

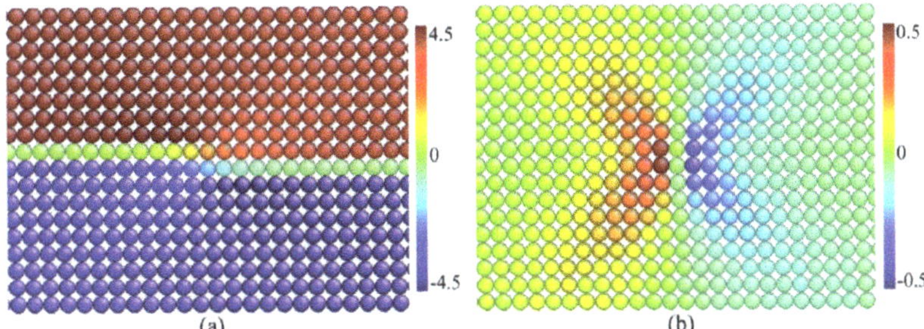

Figure 10. A static kink in a twin wall is shown by (**a**) the vertical shear angle and (**b**) the horizontal shear angle. The strain field in (b) is similar to those of shear dislocations. Reproduced with permission from [105], published by John Wiley & Sons, Inc., 2017.

These kinks generate large strain fields perpendicular and parallel to the twin walls. The kinks move along the twin walls when external stresses are applied. The direction of travel is determined by the energy gain by increasing the energetically favorable twin domain and reducing the unfavorable twin domain. The overall location of the twin wall does not change; only the kink inside the twin wall moves until it hits the sample surface. The self-energy of a moving wall diverges in Landau–Ginzburg theory as $E \sim (1 - v^2/c^2)^{-0.5}$, where c is the relevant sound velocity and v is the wall velocity. This divergence stems from the one-dimensional (1D) character of the wall movement where the propagation direction coincides with the strain gradient as discussed in detail in [25]. Kinks do not suffer from this singularity, because the propagation and strain gradient are rotated by 45° with respect to each other, so that any analytical description is intrinsically 2D avoiding the mass renormalization. Nevertheless, moving kinks also dissipate energy, and this dissipation is the "stumbling block" for high-speed applications. Kinks within the most commonly discussed Φ^4 model (the self-energy of the kink is a 4th order polynomial) contain "wobbles" as internal degrees of freedom [106]. Certain speeds in excess of the sound barrier are theoretically stable and they generate emanating elastic waves during the propagation of the kink [107]. Molecular dynamics (MD) simulations of "realistic" kinks show that the real situation is rather more intriguing: the kink is first accelerated from a static position to a speed near the transverse sound velocity. Further acceleration leads to a maximum velocity, which is greater than the longitudinal sound velocity. Thus, kinks in a planar wall move at ultrasonic speeds if the driving force is sufficiently strong. The kink profiles change during this movement. Phonons are emitted from the moving kink at all velocities while the static kink only induces a strain field that is similar to that of an edge dislocation. When kinks in twin walls propagate with velocities faster than the various speeds of sound in a mineral, they emit secondary waves (~ ultrasonic 'bangs') similar to high-speed projectiles. This effect is speculated to be useful for fast computer memories where the kink position serves as a memory element. Highly twinned leucite is such an example where kinks in twin walls are expected. The first experiments were conducted [108] and they showed that the topological conditions for ultrasonic kinks are favorable. However, the kinks did not contain polarity, or only very weak polarity, so that any electric or magnetic observation of the kink movement would be very difficult.

Magnetic signals of moving kinks in twin walls were computer simulated and they are now predicted for perovskite structures [109]. These results confirm the idea that twins and twin walls commonly produce vortex structures [110,111], as shown in Figure 11.

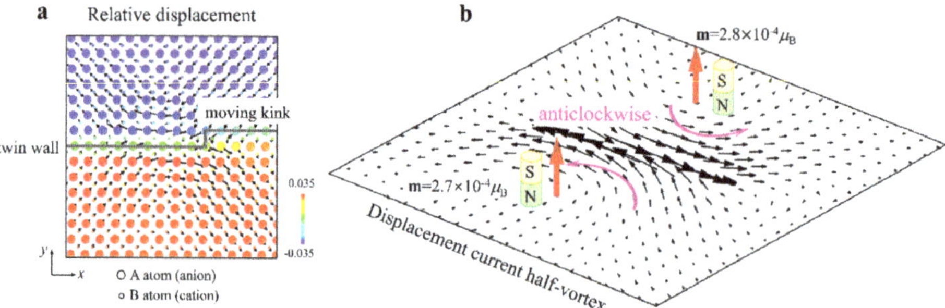

Figure 11. (**a**) Displacement currents and the corresponding magnetic fields produced by moving a kink inside a twin wall. Relative displacements of cations and anions near the moving kink during a time interval of 0.5 ps. The colours are coded by the atomic-level shear strain. Ionic displacements are amplified by a factor of 50 for clarity. (**b**) Two half-vortices of the displacement currents rotate in the same direction in both domains. The current density was calculated based on the relative displacements of anions and cations. The local displacement current is $\sim 10^{-19}$ A. The magnetic field is $\sim 2.7 \times 10^{-4}$ μB for each half-vortex (after [109]).

Twin domains are seen at the surface of the mineral, the walls between twin domains appear as lines, which are often referred to as 'striation'. This leads to the question how the surface properties of a mineral are influenced by the emerging properties of the twin walls. While this problem is greatly discussed in the case of magnetic and ferroelectric materials, much less is known for ferroelastic twin boundaries. The first result is that the surface is bent near the striation line [112,113]. Figure 12 depicts a cut through of the sample perpendicular to the surface. It shows that the surface plane is compressed near the twin boundary, and that long ranging relaxations extend over the surface. These relaxations make the twin boundary appear in optical studies blurred and much wider than their intrinsic geometrical core of some nanometers. The surface forms valleys and ridges where the two twin variants meet at the twin boundary.

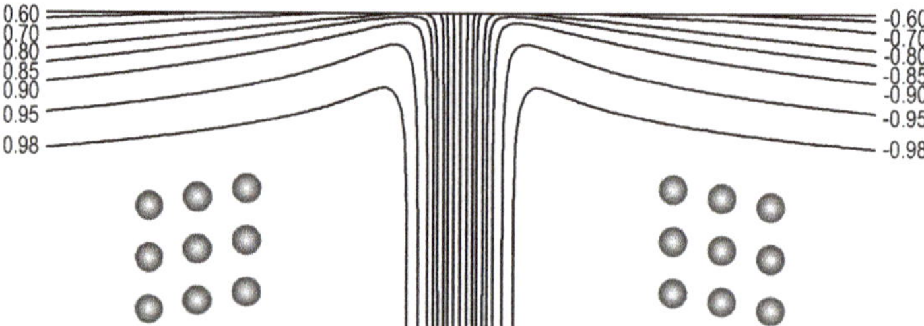

Figure 12. Distribution of the order parameter shear deformation (~order parameter Q) at the surface of the lattice (first 50 layers). Lines represent constant Q, with $Q_0 = 1$ in the bulk. There are three lines in the middle of the twin domain wall that are not labelled, they represent the Q values of 0.40, 0.00, and −0.40 respectively. Notice the steepness of the gradient of Q through the twin domain wall. The two structures represent sheared twin atomic configurations in the bulk (far from the twin domain wall and surfaces). Reproduced with permission from [112], published by IOP Publishing Ltd., 1998.

When the mineral is not mono-atomic, the cations and anions will form dipoles, as discussed above. When these dipoles approach the surface, they are modified by surface relaxations [114], as shown in Figure 13.

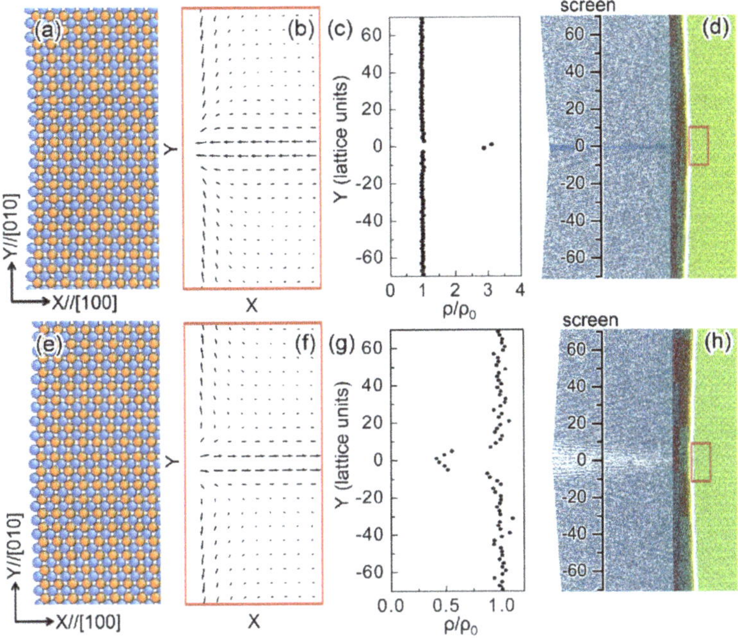

Figure 13. Polarity at anion-terminated surfaces. (**a,b**) The atomic configurations and polarizations near twin boundary I (a valley). The dipole displacement is amplified by a factor of 25 for clarity. Cations are colored in orange and anions are colored in blue. (**c,d**) A snapshot illustrates incident electrons being reflected by the surface. Red electrons are approaching the surface and blue electrons are departing from the surface (the mixture gives rise to the gray color of the electrons). The red rectangle in the green sample indicates the region shown in (a) and (b). The reduced density (ρ/ρ_0) of electrons projected on the virtual screen is shown in (c), ρ_0 is the electron density in the initial configuration. (**e–h**) are the corresponding results for twin boundary II (a ridge). Reproduced with permission from [114]), published by American Physical Society, 2019.

The electron scattering MEM image shown in Figure 14 shows the intricate twin pattern in a perovskite mineral (a) and the equivalent AFM pattern (b). The twin domains show the same surface charges and, hence, no variation in the surface polarization. In contrast, the twin walls show the typical positive or negative surface polarization, which confirms the fact that twin walls in perovskite are polar and piezoelectric.

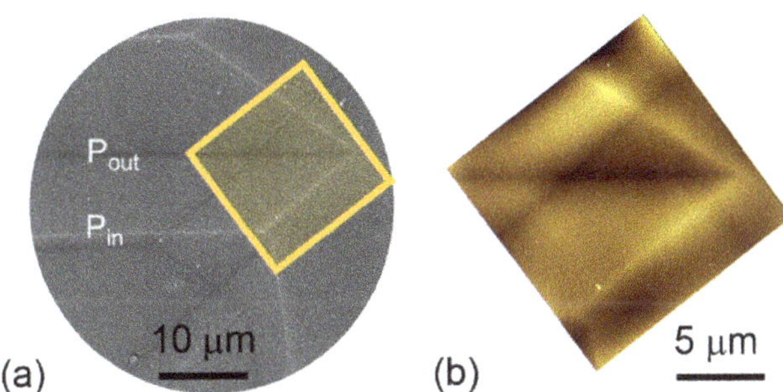

Figure 14. (**a**) Experimental images of electron back scattered MEM image of the $CaTiO_3$ (111) surface showing bright and dark domain walls. Far from the domain wall there is no surface potential contrast between domains. The domain walls are either dark or bright lines, reflecting positive or negative surface topological charge (upward- or downwards-pointing polarity). The orange square is the zone analysed by AFM in (**b**). Reproduced with permission from [114]), published by American Physical Society, 2019.

6. Outlook

Twin structures were well studied in several minerals. Their remarkable feature is that the joint between twins is not simply a 'glue' that holds the twins together, but a highly complex structural unit that differs greatly from the atomic structure of the twin domains. The simple interpolation between atomic positions of adjacent twin domains does not yield the structure of the twin wall. Many exciting properties exist only in the twin wall but not in the twin domains. Some were discussed in this short review, such as chemical storage, polarity, piezoelectricity, topological defects, like kinks, and surface deformations. This observation leads to the question of what happens to twins in minerals that are non-ferroelastic [115]. Growth twins are an excellent candidate for further research. We have no information regarding the atomic structure of such twin walls, but we may speculate that the same principles, which were discussed in this paper, also apply. An atypical example is the twinning in cordierte that contains ferroelastic and topological aspects [116]. This means that these twinned minerals may contain a multitude of 'emerging' properties that we simply do not know. They may contain pockets of material with homeopathic doses of dopants and novel structural elements which are stable only under the geometrical confinement of the twin wall. This has significantly wider consequences: many properties of solids have first been found in minerals before laboratory-based work reproduced the same materials for technological applications. This 'learning from nature' served us well over the last decades. It may well be, I beg to predict, that some of the most exciting properties that are related to twinning are still to be discovered.

Funding: This research was funded by EPSRC, grant number EP/P024904/1 and the EU's Horizon 2020 programme under the Marie Skłodowska-Curie Grant 861153.

Data Availability Statement: Data will be made available under reasonable request.

Conflicts of Interest: The author declares no conflict of interest.

References

1. Kunz, G.F. The Life and Work of Haüy. *Am. Miner.* **1918**, *3*, 59–89.
2. Gruner, J.W. Structural reasons for oriented intergrowths in some minerals. *Am. Miner.* **1929**, *14*, 227–237.
3. Aminoff, G.; Broomé, B. Strukturtheoretische Studien über Zwillinge. I. *Z. Für Krist. Cryst. Mater.* **1931**, *80*, 355–376. [CrossRef]
4. Taylor, W.H.; Darbyshire, J.A.; Strunz, H. An X-ray investigation of the felspars. *Z. Für Krist. Cryst. Mater.* **1934**, *87*, 464–498. [CrossRef]
5. Zachariasen, W.H. The Crystal Structure of Sodium Formate, $NaHCO_2$[1]. *J. Am. Chem. Soc.* **1940**, *62*, 1011–1013. [CrossRef]
6. Friedel, G. *Leçons de Cristallographie*; Berger-Levrault: Paris, France, 1926.
7. Goldsmith, J.R.; Laves, F. Uber die Mischkristallreihe Mikrokline-Albite. *Angew. Chem.* **1956**, *68*, 755–756.
8. Bragg, W.L. *Geological Magazine, London: Humphrey Milford*; Oxford University Press: Oxford, UK, 1937.
9. Strunz, H. Systematik und Struktur der Silikate. *Z. Für Krist. Cryst. Mater.* **1938**, *98*, 60–83. [CrossRef]
10. Klockmann's Lehrbuch der Mineralogie. Paul Ramdohr. *J. Geol.* **1955**, *63*, 99.
11. Aird, A.; Salje, E.K.H. Sheet superconductivity in twin walls: Experimental evidence of. *J. Phys. Condens. Matter* **1998**, *10*, L377. [CrossRef]
12. Aird, A.; Domeneghetti, M.C.; Mazzi, F.; Tazzoli, V.; Salje, E.K.H. Sheet superconductivity in WO_3: Crystal structure of the tetragonal matrix. *J. Phys. Condens. Matter* **1998**, *10*, L569. [CrossRef]
13. Kim, Y.; Alexe, M.; Salje, E.K.H. Nanoscale properties of thin twin walls and surface layers in piezoelectric WO_3-x. *Appl. Phys. Lett.* **2010**, *96*, 032904. [CrossRef]
14. Salje, E.K.H. Multiferroic Domain Boundaries as Active Memory Devices: Trajectories Towards Domain Boundary Engineering. *ChemPhysChem* **2010**, *11*, 940–950. [CrossRef] [PubMed]
15. Salje, E.; Zhang, H. Domain boundary engineering. *Phase Transit.* **2009**, *82*, 452–469. [CrossRef]
16. Nataf, G.F.; Salje, E.K.H. Avalanches in ferroelectric, ferroelastic and coelastic materials: Phase transition, domain switching and propagation. *Ferroelectrics* **2020**, *569*, 82–107. [CrossRef]
17. Nataf, G.F.; Guennou, M.; Gregg, J.M.; Meier, D.; Hlinka, J.; Salje, E.K.H.; Kreisel, J. Domain-wall engineering and topological defects in ferroelectric and ferroelastic materials. *Nat. Rev. Phys.* **2020**, *2*, 634–648. [CrossRef]
18. Carpenter, M.A.; Salje, E.K.H.; Graeme-Barber, A.; Wruck, B.; Dove, M.T.; Knight, K.S. Calibration of excess thermodynamic properties and elastic constant variations associated with the alpha beta phase transition in quartz. *Am. Miner.* **1998**, *83*, 2–22. [CrossRef]

19. Salje, E.K.H.; Ridgwell, A.; Guttler, B.; Wruck, B.; Dove, M.T.; Dolino, G. On the displacive character of the phase transition in quartz: A hard-mode spectroscopy study. *J. Phys. Condens. Matter* **1992**, *4*, 571. [CrossRef]
20. Calleja, M.; Dove, M.T.; Salje, E.K.H. Anisotropic ionic transport in quartz: The effect of twin boundaries. *J. Phys. Condens. Matter* **2001**, *13*, 9445–9454. [CrossRef]
21. Salje, E.; Kuscholke, B.; Wruck, B.; Kroll, H. Thermodynamics of sodium feldspar II: Experimental results and numerical calculations. *Phys. Chem. Miner.* **1985**, *12*, 99–107. [CrossRef]
22. Salje, E. Thermodynamics of sodium feldspar I: Order parameter treatment and strain induced coupling effects. *Phys. Chem. Min.* **1985**, *12*, 93–98. [CrossRef]
23. Tsatskis, I.; Salje, E.K.H. Time evolution of pericline twin domains in alkali feldspars; a computer-simulation study. *Am. Miner.* **1996**, *81*, 800–810. [CrossRef]
24. Salje, E.; Kuscholke, B.; Wruck, B. Domain wall formation in minerals: I. theory of twin boundary shapes in Na-feldspar. *Phys. Chem. Miner.* **1985**, *12*, 132–140. [CrossRef]
25. Salje, E.K. *Phase Transitions in Ferroelastic and Co-Elastic Crystals*; Cambridge University Press: Cambridge, UK, 1993; ISBN 0521429366.
26. Salje, E.K.H. Application of Landau theory for the analysis of phase transitions in minerals. *Phys. Rep.* **1992**, *215*, 49–99. [CrossRef]
27. Carpenter, M.A.; Salje, E.K.H. Elastic anomalies in minerals due to structural phase transitions. *Eur. J. Miner.* **1998**, *10*, 693–812. [CrossRef]
28. Carpenter, M.A.; Salje, E.K.H.; Graeme-Barber, A. Spontaneous strain as a determinant of thermodynamic properties for phase transitions in minerals. *Eur. J. Mineral.* **1998**, *10*, 621–691. [CrossRef]
29. Carpenter, M.A.; Salje, E. Time-dependent Landau theory for order/disorder processes in minerals. *Mineral. Mag.* **1989**, *53*, 483–504. [CrossRef]
30. Sapriel, J. Domain-wall orientations in ferroelastics. *Phys. Rev. B Condens. Matter* **1975**, *12*, 5128–5140. [CrossRef]
31. Yokota, H.; Matsumoto, S.; Salje, E.K.H.; Uesu, Y. Polar nature of domain boundaries in purely ferroelastic $Pb_3(PO_4)_2$ investigated by second harmonic generation microscopy. *Phys. Rev. B* **2019**, *100*, 024101. [CrossRef]
32. Bismayer, U.; Salje, E. Ferroelastic phases in $Pb_3(PO_4)_2$–$Pb_3(AsO_4)_2$; X-ray and optical experiments. *Acta Crystallogr. Sect. A* **1981**, *37*, 145–153. [CrossRef]
33. Hayward, S.A.; Salje, E.K.H. Twin memory and twin amnesia in anorthoclase. *Miner. Mag.* **2000**, *64*, 195–200. [CrossRef]
34. Hayward, S.A.; Salje, E.K.H.; Chrosch, J. Local fluctuations in feldspar frameworks. *Mineral. Mag.* **1998**, *62*, 639–645. [CrossRef]
35. Salje, E.K.H.; Bismayer, U.; Hayward, S.A.; Novak, J. Twin walls and hierarchical mesoscopic structures. *Miner. Mag.* **2000**, *64*, 201–211. [CrossRef]
36. Cámara, F.; Doukhan, J.C.; Salje, E.K.H. Twin walls in anorthoclase are enriched in alkali and depleted in Ca and Al. *Phase Transit.* **2000**, *71*, 227–242. [CrossRef]
37. Aird, A.; Salje, E.K.H. Enhanced reactivity of domain walls in with sodium. *Eur. Phys. J. B Condens. Matter Complex. Syst.* **2000**, *15*, 205–210. [CrossRef]
38. Salje, E.K.H. Fast Ionic Transport Along Twin Walls in Ferroelastic Minerals. In *Properties of Complex Inorganic Solids 2*; Springer: Berlin/Heidelberg, Germany, 2000; pp. 3–15.
39. Yund, R.A.; Quigley, J.; Tullis, J. The effect of dislocations on bulk diffusion in feldspars during metamorphism. *J. Metamorph. Geol.* **1989**, *7*, 337–341. [CrossRef]
40. Hayward, S.A.; Chrosch, J.; Salje, E.K.H.; Carpenter, M.A. Thickness of pericline twin walls in anorthoclase: An X-ray diffraction study. *Eur. J. Mineral.* **1997**, *8*, 1301–1310. [CrossRef]
41. He, X.; Salje, E.K.H.; Ding, X.; Sun, J. Immobile defects in ferroelastic walls: Wall nucleation at defect sites. *Appl. Phys. Lett.* **2018**, *112*, 092904. [CrossRef]
42. Casals, B.; Nataf, G.F.; Salje, E.K.H. Avalanche criticality during ferroelectric/ferroelastic switching. *Nat. Commun.* **2021**, *12*, 345. [CrossRef]
43. Casals, B.; Nataf, G.F.; Pesquera, D.; Salje, E.K.H. Avalanches from charged domain wall motion in $BaTiO_3$ during ferroelectric switching. *Appl. Mater.* **2020**, *8*, 011105. [CrossRef]
44. Salje, E.K.H.; Ding, X.; Zhao, Z.; Lookman, T.; Saxena, A. Thermally activated avalanches: Jamming and the progression of needle domains. *Phys. Rev. B Condens. Matter* **2011**, *83*, 104109. [CrossRef]
45. Harrison, R.J.; Salje, E.K.H. Ferroic switching, avalanches, and the Larkin length: Needle domains in $LaAlO_3$. *Appl. Phys. Lett.* **2011**, *99*, 151915. [CrossRef]
46. Salje, E.K.H.; Dahmen, K.A. Crackling Noise in Disordered Materials. *Annu. Rev. Condens. Matter Phys.* **2014**, *5*, 233–254. [CrossRef]
47. He, X.; Li, S.; Ding, X.; Sun, J.; Selbach, S.M.; Salje, E.K.H. The interaction between vacancies and twin walls, junctions, and kinks, and their mechanical properties in ferroelastic materials. *Acta Mater.* **2019**, *178*, 26–35. [CrossRef]
48. Saidani, S.; Smith, A.; El Hafiane, Y.; Ben Tahar, L. Role of dopants (B, P and S) on the stabilization of β-Ca_2SiO_4. *J. Eur. Ceram. Soc.* **2021**, *41*, 880–891. [CrossRef]
49. Qin, H.; Zhu, J.; Li, N.; Wu, H.; Guo, F.; Sun, S.; Qin, D.; Pennycook, S.J.; Zhang, Q.; Cai, W.; et al. Enhanced mechanical and thermoelectric properties enabled by hierarchical structure in medium-temperature Sb_2Te_3 based alloys. *Nano Energy* **2020**, *78*, 105228. [CrossRef]

50. Boryakov, A.V.; Gladilin, A.A.; Il'ichev, N.N.; Kalinushkin, V.P.; Mironov, S.A.; Rezvanov, R.R.; Uvarov, O.V.; Chegnov, V.P.; Chegnova, O.I.; Chukichev, M.V.; et al. The Influence of Annealing in Zinc Vapor on the Visible and Mid-IR Luminescence of ZnSe:Fe^{2+}. *Opt. Spectrosc.* **2020**, *128*, 1844–1850. [CrossRef]
51. Padrón-Navarta, J.A.; Barou, F.; Daneu, N. Twinning in SnO$_2$-based ceramics doped with CoO and Nb$_2$O$_5$: Morphology of multiple twins revealed by electron backscatter diffraction. *Acta Cryst. B Struct Sci Cryst Eng. Mater.* **2020**, *76*, 875–883. [CrossRef]
52. Ji, X.; Krishnamurthy, M.N.; Lv, D.; Li, J.; Jin, C. Post-synthesis Tellurium Doping Induced Mirror Twin Boundaries in Monolayer Molybdenum Disulfide. *Appl. Sci.* **2020**, *10*, 4758. [CrossRef]
53. Wang, B.; Xia, Y.; Zhang, J.; Komsa, H.-P.; Xie, M.; Peng, Y.; Jin, C. Niobium doping induced mirror twin boundaries in MBE grown WSe$_2$ monolayers. *Nano Res.* **2020**, *13*, 1889–1896. [CrossRef]
54. Yokota, H.; Matsumoto, S.; Hasegawa, N.; Salje, E.K.H.; Uesu, Y. Enhancement of polar nature of domain boundaries in ferroelastic Pb$_3$(PO$_4$)$_2$ by doping divalent-metal ions. *J. Phys. Condens. Matter* **2020**, *32*, 345401. [CrossRef] [PubMed]
55. Calleja, M.; Dove, M.T.; Salje, E.K.H. Trapping of oxygen vacancies on twin walls of CaTiO$_3$: A computer simulation study. *J. Phys. Condens. Matter* **2003**, *15*, 2301–2307. [CrossRef]
56. Goncalves-Ferreira, L.; Redfern, S.A.T.; Artacho, E.; Salje, E.; Lee, W.T. Trapping of oxygen vacancies in the twin walls of perovskite. *Phys. Rev. B* **2010**, *81*, 024109. [CrossRef]
57. Wruck, B.; Salje, E.K.H.; Zhang, M.; Abraham, T.; Bismayer, U. On the thickness of ferroelastic twin walls in lead phosphate Pb$_3$(PO$_4$)$_2$ an X-ray diffraction study. *Phase Transit.* **1994**, *48*, 135–148. [CrossRef]
58. Locherer, K.R.; Chrosch, J.; Salje, E.K.H. Diffuse X-ray scattering in WO$_3$. *Phase Transit.* **1998**, *67*, 51–63. [CrossRef]
59. Chrosch, J.; Salje, E.K.H. Temperature dependence of the domain wall width in LaAlO$_3$. *J. Appl. Phys.* **1999**, *85*, 722–727. [CrossRef]
60. Van Aert, S.; Turner, S.; Delville, R.; Schryvers, D.; Van Tendeloo, G.; Salje, E.K.H. Direct observation of ferrielectricity at ferroelastic domain boundaries in CaTiO$_3$ by electron microscopy. *Adv. Mater.* **2012**, *24*, 523–527. [CrossRef]
61. Salje, E.K.H.; Lee, W.T. Pinning down the thickness of twin walls. *Nat. Mater.* **2004**, *3*, 425–426. [CrossRef]
62. Lee, W.T.; Salje, E.K.H.; Bismayer, U. Surface relaxations at mineral surfaces. *Z. Für Krist. Cryst. Mater.* **2005**, *220*, 683–690. [CrossRef]
63. Lee, W.T.; Salje, E.K.H. Chemical turnstile. *Appl. Phys. Lett.* **2005**, *87*, 143110. [CrossRef]
64. Shilo, D.; Mendelovich, A.; Novák, V. Investigation of twin boundary thickness and energy in CuAlNi shape memory alloy. *Appl. Phys. Lett.* **2007**, *90*, 193113. [CrossRef]
65. Shilo, D.; Ravichandran, G.; Bhattacharya, K. Investigation of twin-wall structure at the nanometre scale using atomic force microscopy. *Nat. Mater.* **2004**, *3*, 453–457. [CrossRef] [PubMed]
66. Houchmandzadeh, B.; Lajzerowicz, J.; Salje, E. Order parameter coupling and chirality of domain walls. *J. Phys. Condens. Matter* **1991**, *3*, 5163–5169. [CrossRef]
67. Houchmanzadeh, B.; Lajzerowicz, J.; Salje, E. Interfaces and ripple states in ferroelastic crystals—A simple model. *Phase Transit.* **1992**, *38*, 77–87. [CrossRef]
68. Houchmandzadeh, B.; Lajzerowicz, J.; Salje, E. Relaxations near surfaces and interfaces for first-, second- and third-neighbour interactions: Theory and applications to polytypism. *J. Phys. Condens. Matter* **1992**, *4*, 9779–9794. [CrossRef]
69. Salje, E.K.H. Ferroelastic domain walls as templates for multiferroic devices. *J. Appl. Phys.* **2020**, *128*, 164104. [CrossRef]
70. Salje, E.K.H.; Aktas, O.; Carpenter, M.A.; Laguta, V.V.; Scott, J.F. Domains within domains and walls within walls: Evidence for polar domains in cryogenic SrTiO$_3$. *Phys. Rev. Lett.* **2013**, *111*, 247603. [CrossRef]
71. Aktas, O.; Salje, E.K.H.; Crossley, S.; Lampronti, G.I.; Whatmore, R.W.; Mathur, N.D.; Carpenter, M.A. Ferroelectric precursor behavior in PbSc$_{0.5}$Ta$_{0.5}$O$_3$ detected by field-induced resonant piezoelectric spectroscopy. *Phys. Rev. B* **2013**, *88*, 174112. [CrossRef]
72. Viehland, D.D.; Salje, E.K.H. Domain boundary-dominated systems: Adaptive structures and functional twin boundaries. *Adv. Phys.* **2014**, *63*, 267–326. [CrossRef]
73. Lee, W.T.; Salje, E.K.H.; Goncalves-Ferreira, L.; Daraktchiev, M.; Bismayer, U. Intrinsic activation energy for twin-wall motion in the ferroelastic perovskite CaTiO$_3$. *Phys. Rev. B Condens. Matter* **2006**, *73*, 214110. [CrossRef]
74. Yokota, H.; Usami, H.; Haumont, R.; Hicher, P.; Kaneshiro, J.; Salje, E.K.H.; Uesu, Y. Direct evidence of polar nature of ferroelastic twin boundaries inCaTiO$_3$ obtained by second harmonic generation microscope. *Phys. Rev. B* **2014**, *89*, 144109. [CrossRef]
75. Salje, E.K.H. Hard mode Spectroscopy: Experimental studies of structural phase transitions. *Phase Transit.* **1992**, *37*, 83–110. [CrossRef]
76. Nataf, G.F.; Guennou, M. Optical studies of ferroelectric and ferroelastic domain walls. *J. Phys. Condens. Matter* **2020**, *32*, 183001. [CrossRef]
77. Yokota, H.; Hasegawa, N.; Glazer, M.; Salje, E.K.H.; Uesu, Y. Direct evidence of polar ferroelastic domain boundaries in semiconductor BiVO$_4$. *Appl. Phys. Lett.* **2020**, *116*, 232901. [CrossRef]
78. Bridge, P.J.; Pryce, M.W. Clinobisvanite, monoclinic BiVO$_4$, a new mineral from Yinnietharra, Western Australia. *Mineral. Mag.* **1974**, *39*, 847–849. [CrossRef]
79. Yokota, H.; Haines, C.R.S.; Matsumoto, S.; Hasegawa, N.; Carpenter, M.A.; Heo, Y.; Marin, A.; Salje, E.K.H.; Uesu, Y. Domain wall generated polarity in ferroelastics: Results from resonance piezoelectric spectroscopy, piezoelectric force microscopy, and optical second harmonic generation measurements in LaAlO$_3$ with twin and tweed microstructures. *Phys. Rev. B* **2020**, *102*, 104117. [CrossRef]

90. Salje, E.; Parlinski, K. Microstructures in high Tc superconductors. *Supercond. Sci. Technol.* **1991**, *4*, 93. [CrossRef]
91. Parlinski, K.; Heine, V.; Salje, E.K.H. Origin of tweed texture in the simulation of a cuprate superconductor. *J. Phys. Condens. Matter* **1993**, *5*, 497–518. [CrossRef]
92. Wang, X.; Salje, E.K.H.; Sun, J.; Ding, X. Glassy behavior and dynamic tweed in defect-free multiferroics. *Appl. Phys. Lett.* **2018**, *112*, 012901. [CrossRef]
93. Bratkovsky, A.M.; Marais, S.C.; Heine, V.; Salje, E.K.H. The theory of fluctuations and texture embryos in structural phase transitions mediated by strain. *J. Phys. Condens. Matter* **1994**, *6*, 3679. [CrossRef]
94. Putnis, A.; Salje, E. Tweed microstructures: Experimental observations and some theoretical models. *Phase Transit.* **1994**, *48*, 85–105. [CrossRef]
95. Bratkovsky, A.M.; Salje, E.K.H.; Heine, V. Overview of the origin of tweed texture. *Phase Transitions* **1994**, *52*, 77–83. [CrossRef]
96. Salje, E.K.H. Tweed, twins, and holes. *Am. Miner.* **2015**, *100*, 343–351. [CrossRef]
97. Tribaudino, M.; Benna, P.; Bruno, E. I 1-I 2/c phase transition in alkaline-earth feldspars: Evidence from TEM observations of Sr-rich feldspar along the $CaAl_2Si_2O_8$-$SrAl_2Si_2O_8$ join. *Am. Miner.* **1995**, *80*, 907–915. [CrossRef]
98. Janney, D.E.; Wenk, H.-R. Peristerite exsolution in metamorphic plagioclase from the Lepontine Alps: An analytical and transmission electron microscope study. *Am. Miner.* **1999**, *84*, 517–527. [CrossRef]
99. McGuinn, M.D.; Redfern, S.A.T. High-temperature ferroelastic strain below the I2/c-I1 transition in $Ca_{1-x}Sr_xAl_2Si_2O_8$ feldspars. *Eur. J. Miner.* **1997**, *9*, 1159–1172. [CrossRef]
100. Viehland, D.; Dai, X.H.; Li, J.F.; Xu, Z. Effects of quenched disorder on La-modified lead zirconate titanate: Long- and short-range ordered structurally incommensurate phases, and glassy polar clusters. *J. Appl. Phys.* **1998**, *84*, 458–471. [CrossRef]
101. Lee, M.R.; Hodson, M.E.; Parsons, I. The role of intragranular microtextures and microstructures in chemical and mechanical weathering: Direct comparisons of experimentally and naturally weathered alkali feldspars. *Geochim. Cosmochim. Acta* **1998**, *62*, 2771–2788. [CrossRef]
102. Lee, M.R.; Brown, D.J.; Smith, C.L.; Hodson, M.E.; MacKenzie, M.; Hellmann, R. Characterization of mineral surfaces using FIB and TEM: A case study of naturally weathered alkali feldspars. *Am. Miner.* **2007**, *92*, 1383–1394. [CrossRef]
103. Carpenter, M.A.; Zhang, Z. Anelasticity maps for acoustic dissipation associated with phase transitions in minerals. *Geophys. J. Int.* **2011**, *186*, 279–295. [CrossRef]
104. Carpenter, M.A. Static and dynamic strain coupling behaviour of ferroic and multiferroic perovskites from resonant ultrasound spectroscopy. *J. Phys. Condens. Matter* **2015**, *27*, 263201. [CrossRef]
105. Perks, N.J.; Zhang, Z.; Harrison, R.J.; Carpenter, M.A. Strain relaxation mechanisms of elastic softening and twin wall freezing associated with structural phase transitions in $(Ca,Sr)TiO_3$ perovskites. *J. Phys. Condens. Matter* **2014**, *26*, 505402. [CrossRef] [PubMed]
106. Oravova, L.; Zhang, Z.; Church, N.; Harrison, R.J.; Howard, C.J.; Carpenter, M.A. Elastic and anelastic relaxations accompanying magnetic ordering and spin-flop transitions in hematite, Fe_2O_3. *J. Phys. Condens. Matter* **2013**, *25*, 116006. [CrossRef]
107. Zhang, Z.; Koppensteiner, J.; Schranz, W.; Carpenter, M.A. Variations in elastic and anelastic properties of Co_3O_4 due to magnetic and spin-state transitions. *Am. Mineral.* **2012**, *97*, 399–406. [CrossRef]
108. Zhang, Z.; Schranz, W.; Carpenter, M.A. Acoustic attenuation due to transformation twins in $CaCl_2$: Analogue behaviour for stishovite. *Phys. Earth Planet. Inter.* **2012**, *206–207*, 43–50. [CrossRef]
109. Carpenter, M.A.; Salje, E.K.H.; Howard, C.J. Magnetoelastic coupling and multiferroic ferroelastic/magnetic phase transitions in the perovskite $KMnF_3$. *Phys. Rev. B Condens. Matter Mater. Phys.* **2012**, *85*, 224430. [CrossRef]
100. Salje, E.K.H.; Zhao, Z.; Ding, X.; Sun, J. Mechanical spectroscopy in twinned minerals: Simulation of resonance patterns at high frequencies. *Am. Miner.* **2013**, *98*, 1449–1458. [CrossRef]
101. Zhao, Z.; Ding, X.; Lookman, T.; Sun, J.; Salje, E.K.H. Mechanical loss in multiferroic materials at high frequencies: Friction and the evolution of ferroelastic microstructures. *Adv. Mater.* **2013**, *25*, 3244–3248. [CrossRef] [PubMed]
102. Salje, E.K.H.; Ding, X.; Zhao, Z.; Lookman, T. How to generate high twin densities in nano-ferroics: Thermal quench and low temperature shear. *Appl. Phys. Lett.* **2012**, *100*, 222905. [CrossRef]
103. Kustov, S.; Liubimova, I.; Salje, E.K.H. Domain Dynamics in Quantum-Paraelectric $SrTiO_3$. *Phys. Rev. Lett.* **2020**, *124*, 016801. [CrossRef] [PubMed]
104. Kustov, S.; Liubimova, I.; Salje, E.K.H. $LaAlO_3$: A substrate material with unusual ferroelastic properties. *Appl. Phys. Lett.* **2018**, *112*, 042902. [CrossRef]
105. Salje, E.K.H.; Wang, X.; Ding, X.; Scott, J.F. Ultrafast switching in avalanche-driven ferroelectrics by supersonic kink movements. *Adv. Funct. Mater.* **2017**, *27*, 1700367. [CrossRef]
106. Kälbermann, G. The sine-Gordon wobble. *J. Phys. A Math. Gen.* **2004**, *37*, 11603–11612. [CrossRef]
107. Barashenkov, I.V.; Oxtoby, O.F. Wobbling kinks in Φ^4 theory. *Phys. Rev. E* **2009**, *80*, 026608. [CrossRef] [PubMed]
108. Aktas, O.; Carpenter, M.A.; Salje, E.K.H. Elastic softening of leucite and the lack of polar domain boundaries. *Am. Miner.* **2015**, *100*, 2159–2162. [CrossRef]
109. Lu, G.; Li, S.; Ding, X.; Sun, J.; Salje, E.K.H. Current vortices and magnetic fields driven by moving polar twin boundaries in ferroelastic materials. *NPJ Comput. Mater.* **2020**, *6*, 1–6. [CrossRef]
110. Salje, E.K.H.; Li, S.; Stengel, M.; Gumbsch, P.; Ding, X. Flexoelectricity and the polarity of complex ferroelastic twin patterns. *Phys. Rev. B Condens. Matter* **2016**, *94*, 024114. [CrossRef]

111. Zhao, Z.; Ding, X.; Salje, E.K.H. Flicker vortex structures in multiferroic materials. *Appl. Phys. Lett.* **2014**, *105*, 112906. [CrossRef]
112. Novak, J.; Salje, E.K.H. Surface structure of domain walls. *J. Phys. Condens. Matter* **1998**, *10*, L359–L366. [CrossRef]
113. Conti, S.; Weikard, U. Interaction between free boundaries and domain walls in ferroelastics. *Eur. Phys. J. B Condens. Matte Complex. Syst.* **2004**, *41*, 413–420. [CrossRef]
114. Zhao, Z.; Barrett, N.; Wu, Q.; Martinotti, D.; Tortech, L.; Haumont, R.; Pellen, M.; Salje, E.K.H. Interaction of low-energy electron with surface polarity near ferroelastic domain boundaries. *Phys. Rev. Mater.* **2019**, *3*, 043601. [CrossRef]
115. Salje, E.K.H. Ferroelastic materials. *Annu. Rev. Mater. Res.* **2012**, *42*, 265–283. [CrossRef]
116. Blackburn, J.F.; Salje, E.K.H. Time evolution of twin domains in cordierite: A computer simulation study. *Phys. Chem. Miner.* **1999**, *26*, 275–291. [CrossRef]

Article

Diffraction Features from (10$\bar{1}$4) Calcite Twins Mimicking Crystallographic Ordering

Péter Németh [1,2]

[1] Research Centre for Astronomy and Earth Sciences, Institute for Geological and Geochemical Research, Eötvös Loránd Research Network, Budaörsi Street 45, 1112 Budapest, Hungary; nemeth.peter@csfk.org

[2] Department of Earth and Environmental Sciences, University of Pannonia, Egyetem út 10, 8200 Veszprém, Hungary

Abstract: During phase transitions the ordering of cations and/or anions along specific crystallographic directions can take place. As a result, extra reflections may occur in diffraction patterns, which can indicate cell doubling and the reduction of the crystallographic symmetry. However, similar features may also arise from twinning. Here the nanostructures of a glendonite, a calcite ($CaCO_3$) pseudomorph after ikaite ($CaCO_3 \cdot 6H_2O$), from Victoria Cave (Russia) were studied using transmission electron microscopy (TEM). This paper demonstrates the occurrence of extra reflections at positions halfway between the Bragg reflections of calcite in *0kl* electron diffraction patterns and the doubling of d_{104} spacings (corresponding to 2·3.03 Å) in high-resolution TEM images. Interestingly, these diffraction features match with the so-called carbonate *c-type* reflections, which are associated with Mg and Ca ordering, a phenomenon that cannot occur in pure calcite. TEM and crystallographic analysis suggests that, in fact, (10$\bar{1}$4) calcite twins and the orientation change of CO_3 groups across the twin interface are responsible for the extra reflections.

Keywords: electron diffraction; *c-type* reflections; ordering; twinning; calcite; glendonite; TEM

Citation: Németh, P. Diffraction Features from (10$\bar{1}$4) Calcite Twins Mimicking Crystallographic Ordering. *Minerals* **2021**, *11*, 720. https://doi.org/10.3390/min11070720

Academic Editor: Thomas N. Kerestedjian

Received: 3 June 2021
Accepted: 1 July 2021
Published: 4 July 2021

Publisher's Note: MDPI stays neutral with regard to jurisdictional claims in published maps and institutional affiliations.

1. Introduction

Transmission electron microscopy (TEM) is an excellent method for characterizing the nanostructures of crystalline materials. A wide range of imaging and diffraction techniques provide unique information about the crystal structure of the studied samples, such as the local elemental composition, electronic structure and bonding down to the atomic level [1]. Thanks to the latest developments, not only has the resolution limit been pushed below 0.5 Å, but nowadays even the behavior of materials and devices at the atomic scale can be mapped [2,3]. TEM is practically an indispensable technique for investigating crystal defects and studying phase transitions. However, the interpretation of TEM data can be challenging, and ambiguities can arise for cases that look simple [4]. In particular, the recognition of twins from TEM data can be difficult because their diffraction features can be confused with ordering and superstructures [5–7].

Here TEM is used to study the nanostructure of glendonite, which is a synonym for calcite pseudomorph after ikaite ($CaCO_3 \cdot 6H_2O$), as a result of transformation. Glendonite is commonly associated with cold paleotemperature [8–10]. However, issues have been raised with regard to using it as an indicator for cold temperature [11] because, in a laboratory setting, ikaite formation was reported even above 20 °C [11,12]. Ikaite rapidly disintegrates into a mush of water and recrystallizes to calcite during slight warming or pressure release, resulting in a highly porous crystal mesh [13–16]. Aragonite (unit cell parameters a = 4.9611 Å, b = 7.9672 Å, c = 5.7404 Å; space group: *Pmcn*) has also been observed from ikaite transformation in marine and alkaline lake environments [17,18]. Studies on the transformation of synthetic ikaite suggest that it can also transform to amorphous calcium carbonate [19] and vaterite (unit cell parameters a = 4.13 Å, b = 7.15 Å, c = 8.48 Å; space group: *Pbnm*) (e.g., [20]). Macroscopic calcite pseudomorphs after ikaite are traditionally called glendonite, although other names also are used [8]. According to [11], the pseudomorph replacement of ikaite by calcite occurs through a coupled dissolution–reprecipitation mechanism at the ikaite–calcite interface. Since glendonites preserved the parent phase morphology, structural relics indicating phase transition can be expected [21]. In fact, the TEM data here reported indicated an interesting lamellar structure with the occurrence of diffraction features that are inconsistent with ordinary calcite and that could mistakenly be associated with crystallographic ordering and superstructures.

Understanding the structural details of Ca-Mg carbonates is an important topic since they are rock-forming minerals. In particular, calcite is the dominant carbonate phase on the Earth's surface and it is the third most abundant mineral after quartz and feldspars. It abundantly precipitates at ambient conditions and is one of the most important biominerals. Calcite is rhombohedral (unit cell parameters a = 4.989 Å and c = 17.06 Å; space group: $R\bar{3}c$). In this paper, three independent indices hkl are used to label reflections and d-spacings and four indices $hkil$ ($i = -(h + k)$) are used whenever crystallographic planes and directions are discussed. Ca atoms lie in the $(10\bar{1}4)$ plane alternating with CO_3 triangular groups along the [0001] (Figure 1a). The orientation change of the carbonate groups implies a c glide that is revealed by systematic absences ($l = 2n + 1$ for $0kl$ reflections) in the corresponding reciprocal lattice (Figure 1b). In dolomite, $(10\bar{1}4)$ Mg and Ca cation layers alternate along [0001]; as a result, the $R\bar{3}c$ symmetry is reduced to $R\bar{3}$ and b-type reflections ($l = 2n + 1$ for $0kl$ reflections) occur (e.g., [22]). Although such b-type reflections do not occur in pure conventional calcite, they are present in a disordered phase of calcite (space group: $R\bar{3}m$) as a consequence of the rotational disorder of the CO_3 groups. This polymorph was reported at high temperature (>1260 K) and at high pCO$_2$ [23]. The reflections halfway between those of the a-type ($l = 2n$ for the $0kl$ reflections) and b-type are the so-called c-type reflections (Figure 1c), which have been associated with ordering and attributed to various superstructures in Mg-bearing calcite and dolomite [22,24,25]. In contrast to these explanations, Larsson and Christy [5] showed that the c-type reflections can arise from submicron-sized calcite twins. Shen et al. [26] followed this explanation and demonstrated that multiple diffraction between the host dolomite and twinned Mg-calcite nano-lamellae could give rise to the c-type reflections in Ca-rich dolomite.

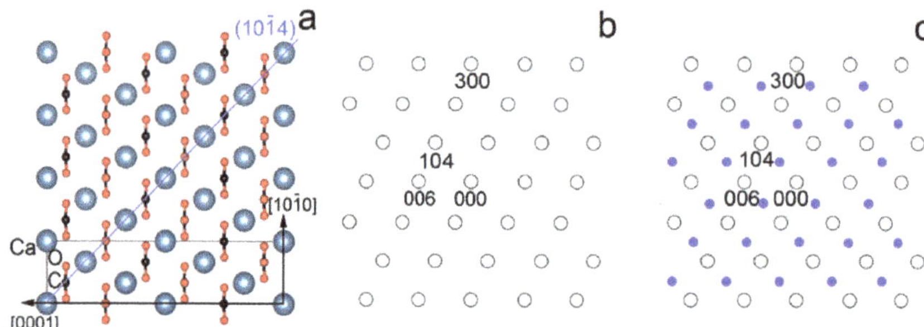

Figure 1. (**a**) Structure of calcite seen along $[01\bar{1}0]$. In order, the large blue circles and red and black dots represent Ca, O and C atoms. The $(10\bar{1}4)$ plane is marked with a blue line. (**b**) $0kl$ diffraction plane of $R\bar{3}c$ calcite with a-type reflections (empty circles). (**c**) $0kl$ diffraction plane with c-type reflections (blue dots) occurring halfway between a-type reflections.

The aim of this paper is to confirm Larsson and Christy's [5] hypothesis via the recognition of calcite $(10\bar{1}4)$ twins from selected-area electron diffraction (SAED) patterns and high-resolution TEM images (HRTEM). Here it is demonstrated that the c-type reflections in glendonite can be successfully interpreted in terms of $(10\bar{1}4)$ twinning by reticular pseudo-merohedry [27]. It is also shown that streaked c-type reflections can be associated with $(10\bar{1}4)$ stacking faults occurring inside the small (5–15 nm wide) twin domains. The proper interpretation of these diffraction features is crucial as they can lead to erroneous conclusions.

2. Materials and Methods

Glendonite samples from Victoria Cave (Bashkiria, Russia) were received from Paul Töchterle and Yuri Dublyansky (Institute of Geology, University of Innsbruck). The description of the site and the characteristics of the samples are reported in [28] and [21] respectively.

Glendonite grains were coated in gold and measured using a Zeiss EVO 40 scanning electron microscope operated at 20 kV. A back-scattered electron (BSE) image was acquired at 10,000 magnifications and the chemical composition of the sample was measured using an Oxford INCA energy dispersive X-ray (EDX) spectrometer.

Glendonite powder was analyzed with an X-ray diffraction (X'Pert, PANalytical B.V.) method with Cu Kα radiation in the 2θ range of 10 to 60°and with a 0.04°step scan for 1 sec. Si was used as an internal standard. The unit-cell parameter of calcite was refined based on the 012, 104, 110, 113, 202, 024 and 116 reflections using the PDIndexer program [29].

Zoltán May (ELKH Research Centre for Natural Sciences, Budapest) determined the elemental composition of glendonite by atomic emission spectroscopy using a Spectro Genesis ICP-OES (SPECTRO Analytical Instruments GmbH, Kleve, Germany) simultaneous spectrometer with axial plasma observation. Multi-element standard solutions for ICP (Merck Chemicals GmbH, Darmstadt, Germany) were used for calibration. The detection limits of the analytical method are reported in [30].

For TEM investigation the samples were crushed in ethanol and deposited onto copper grids covered by Lacey carbon supporting films. Bright-field TEM (BFTEM), HRTEM and SAED data were acquired with a 200 keV Talos Thermo Scientific electron microscope. Fast Fourier transforms (FFTs) obtained from the HRTEM images were calculated using Gatan Digital Micrograph 3.6.1 software. Structure models of calcite were drawn using VESTA-win64 software [31] and the twinning features of calcite were tested with the software GEMINOGRAPHY [32].

3. Results and Discussion

3.1. c-Type Reflections and the Twin Lattice

The studied glendonite is built up of fine-grained (submicrons to several microns in size) calcite crystals (Figure 2a). The sample is chemically homogeneous, practically pure $CaCO_3$ (Figure 2a,b). Its water content is negligible [21] and its X-ray diffraction pattern is consistent with ordinary calcite, having the unit cell parameters a = 4.989 (2) and c = 17.08 (1) (Figure 2c). However, TEM reveals that the sample has a complex nanostructure. In particular, BFTEM images taken along [01$\bar{1}$0] show linear features parallel to (10$\bar{1}$4) and indicate the occurrence of 10–15 nm size nano-domains in the calcite matrix (Figure 3a). SAED patterns of the nano-domains reveal weak *c-type* reflections that are incompatible with the $R\bar{3}c$ space group of calcite (Figure 3b).

It is intriguing that the *c-type* reflections of the studied sample mimic the diffraction features of ikaite (a = 8.792 Å, b = 8.310 Å, c = 11.021 Å, β = 110.53°, space group: $C2/c$) and also those of a hypothetical carbonate supercell containing ordered cation sites. In fact, the measured d spacings of 6.05 Å and 3.85 Å roughly match the d_{110} (5.849 Å) and $d_{2\text{-}1\text{-}1}$ (3.882 Å) of ikaite. In addition, the calculated angle between (110) and (2$\bar{1}\bar{1}$) is 75.7°, which is close to the measured value of 73°. However, ikaite loses its water content in vacuum within seconds; thus, it cannot be studied with TEM. Furthermore, Fourier transform infrared spectroscopy does not indicate a detectable amount (<1 wt%) of crystalline water inside the sample [21]. Similarly, the interpretation of *c-type* reflections with cation ordering is not viable, because both EDX and ICP-OES analysis (Figure 2) suggest that the sample does not contain a detectable amount (<1 wt%) of foreign cations.

In contrast to the abovementioned hypothetical explanations, (10$\bar{1}$4) calcite twins provide a convincing solution for the diffraction features of the SAED pattern (Figure 3b). In particular, twin individuals hosted in an underlying calcite lattice [5] give rise to the observed extra and split reflections (Figure 3c). According to tests using the software GEMINOGRAPHY [32], the observed twin corresponds to a (10$\bar{1}$4) twin by reticular pseudo-merohedry (twin index 4 and obliquity 0.74°) based on a pseudo-orthorhombic C-centered sublattice with cell parameters a = 8.095 Å, b = 24.286 Å, c = 4.989 Å, α = 90.00°, β = 90.00° and γ = 89.26°. The calculated obliquity (0.74°) is in accordance with the small observed split of reflections in the SAED pattern (Figure 3c). The relationship between the unit cell of calcite and that of the pseudo-orthorhombic twin lattice is described by the following matrix T:

$$\begin{pmatrix} 4/3 & 2/3 & -1/3 \\ 4 & 2 & 1 \\ 0 & -1 & 0 \end{pmatrix}$$

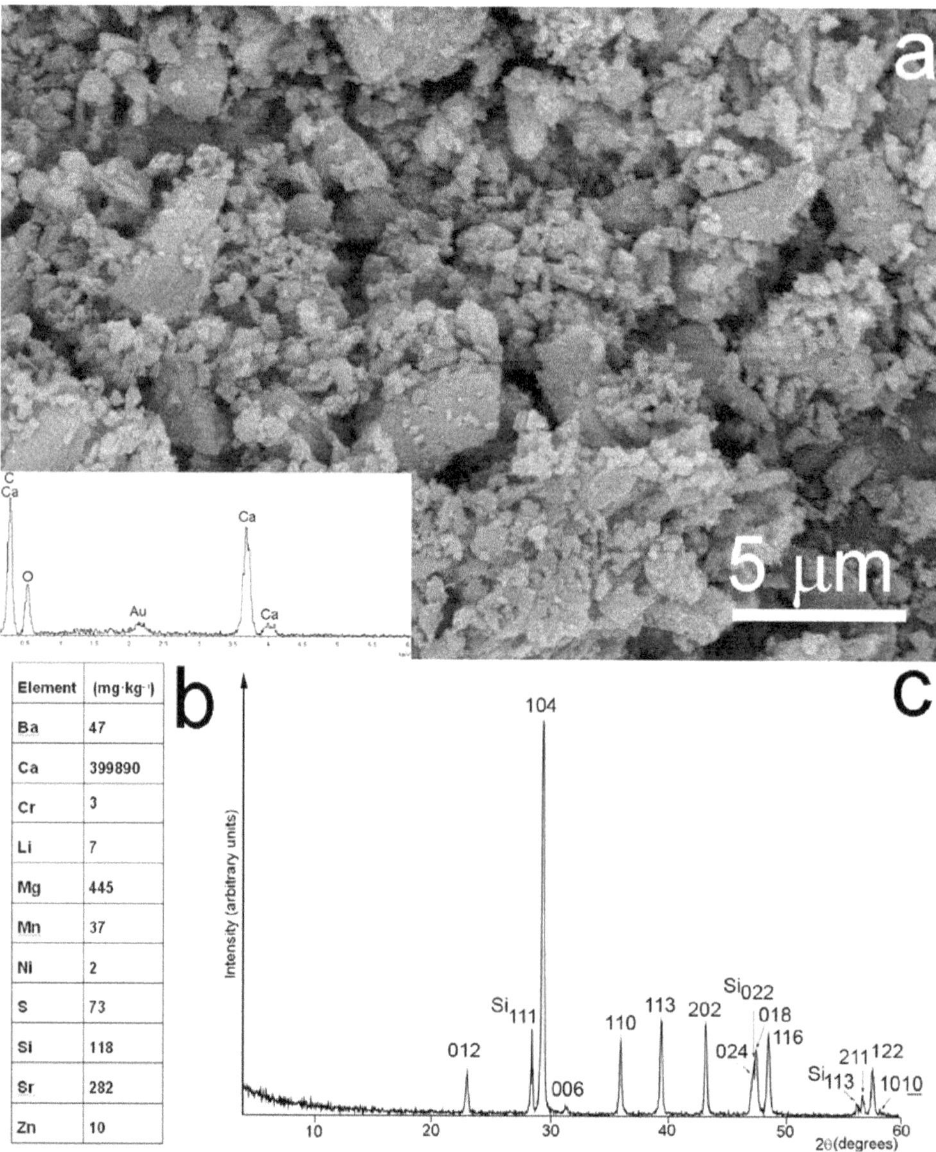

Figure 2. (**a**) BSE image and EDX data of the studied glendonite. (**b**) Elements present in glendonite. (**c**) X-ray powder diffraction pattern of glendonite calcite. Si standard was added for measuring the cell parameters.

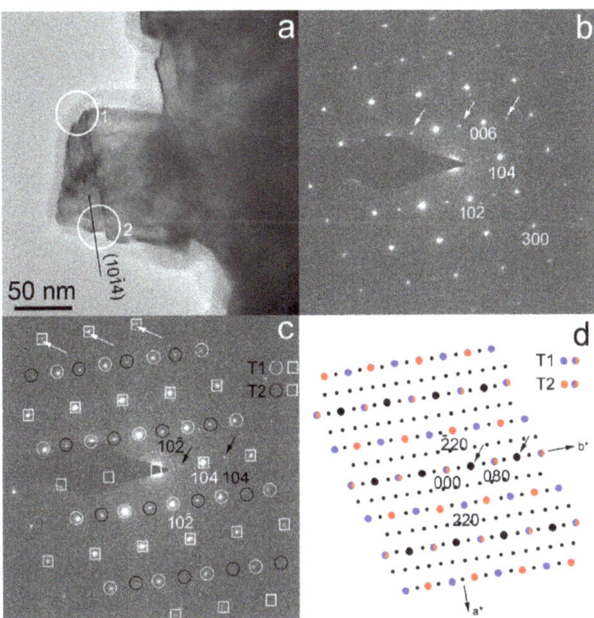

Figure 3. (**a**) BFTEM image of a ~10 nm size (10$\bar{1}$4) twinned domain within the calcite matrix. The areas 1 and 2, marked by white circles, are magnified in Figures 4 and 5, respectively. (**b**) The SAED pattern along [01$\bar{1}$0] obtained from a ~200 nm size region shown in (**a**) reveals weak *c-type* reflections (white arrows). (**c**) The interpretation of the SAED pattern (**b**) via (10$\bar{1}$4) twinning. The splitting of the overlapping reflections (white squares) increases with increasing diffraction angle (white arrows). Black arrows mark reflections arising from dynamically scattered electrons of the (10$\bar{1}$4) twin hosted within a calcite matrix. (**d**) The interpretation of the SAED pattern (**b**) with the pseudo-orthorhombic twin cell (*hk*0 diffraction plane). Large-size black dots denote the reflections marked by black arrows in (**c**).

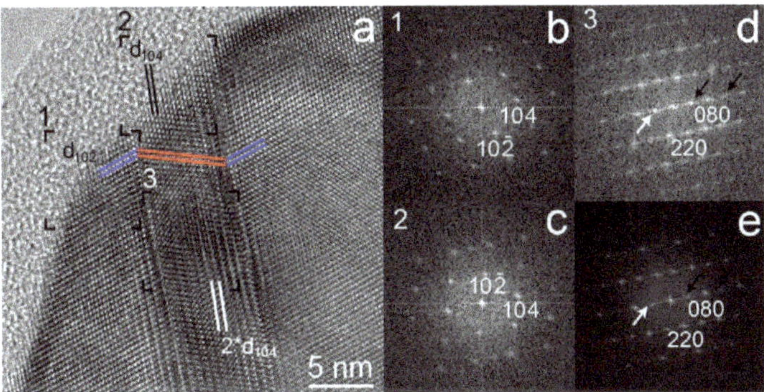

Figure 4. (**a**) HRTEM image of a ~10 nm size (10$\bar{1}$4) twinned domain within calcite along [01$\bar{1}$0] (area shown in Figure 3a with white circle 1). Blue and red lines show $d_{10\text{-}2}$ spacings of three polysynthetic twinned individuals. Black and white lines show single and doubled d_{104} spacings, respectively. FFTs calculated from regions 1 (**b**) and 2 (**c**) correspond to twin individuals 1 and 2. (**d**) FFT calculated from region 3 corresponds to an area where individual 2 is hosted in the calcite matrix. (**e**) FFT calculated from the entire HRTEM image (**a**) shows a diffraction pattern similar to inset (**d**). Dynamically scattered electrons arising from the thick part of the sample result in reflections marked by black arrows in (**d**) and (**e**). The FFTs of (**d**) and (**e**) are interpreted with the pseudo-orthorhombic twin cell. White arrows mark the streaked reflections for (**d**) and (**e**), which are indicative for (10$\bar{1}$4) stacking faults.

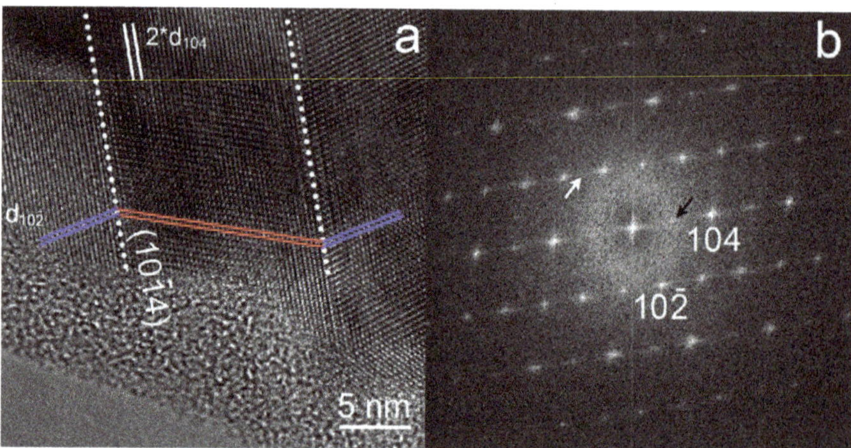

Figure 5. (**a**) HRTEM image of a ~15 nm size (10$\overline{1}$4) twin domain within calcite along [01$\overline{1}$0] (area shown in Figure 3a with white circle 2). Blue and red lines mark $d_{10\text{-}2}$ spacings of the twin individuals. White lines mark doubled d_{104} spacings. (**b**) FFT calculated from the HRTEM image (**a**) showing *c-type* reflections (black arrow). The FFT is indexed according to rhombohedral calcite. The white arrow shows streaked reflections, which can indicate (10$\overline{1}$4) stacking faults.

Applying this matrix to any [u, v, w] vectors of calcite, the corresponding [u′, v′, w′] vectors of the pseudo-orthorhombic cell can be generated. In addition, applying this matrix to indexes of the calcite *hkl* reflections, we can transform them to the *h′k′l′* indexes referred to the twin lattice. For example, the 104 reflection of calcite corresponds to 080 of the pseudo-orthorhombic twin lattice. The twin lattice explains all the observed reflections and shows that every fourth set of reflections of the individuals overlaps on the *hk0* plane (Figure 3d), as expected from the twin index 4.

3.2. HRTEM Images of the (10$\overline{1}$4) Calcite Twins and Doubled d_{104} Spacings

The identification of the (10$\overline{1}$4) calcite twins is straightforward from the HRTEM image acquired with the electron beam incident along [01$\overline{1}$0] of the thin part (~10 nm thick) of the sample; in fact, the two individuals can be easily recognized by the orientation change of the fringes corresponding to d_{012} spacings (3.85 Å) and they can be localized from their corresponding FFTs (Figure 4a–c). The HRTEM image (Figure 4a) obtained from the thin part also shows fringes with 3.03 Å corresponding to d_{104} spacing. However, 6.05 Å spacing can be measured from the thicker areas (>10 nm) of the image (white lines in Figure 4a) and its corresponding FFT (Figure 4d), as well as the FFT calculated from the entire HRTEM image (Figure 4e). This *d*-spacing, which is also evident in Figure 5a, erroneously indicates cell doubling.

It should be noted that the systematic absences of the $R\overline{3}c$ space group, i.e., those of the twin individuals, should be present even for thick samples. Thus, the side-by-side arrangement of the individuals does not provide an explanation for the *c-type* reflections of the FFT calculated from the thick part of the grain (Figure 4d). Therefore, following the suggestion of Larsson and Christy [5], it is proposed that the small (<10 nm) twin individual 2 is actually hosted in the calcite matrix (individual 1). Since HRTEM images are two-dimensional projections of the three-dimensional structures, the HRTEM image acquired with the electron beam incident along [01$\overline{1}$0] shows the vertical projection of the individual 2 and that of the underlying calcite (individual 1) from areas thicker than the individual twin size (>10 nm). It is suggested that electrons dynamically scattered from these vertically stacked twin individuals give rise to the doubled d_{104} spacings (2·3.03 Å) of the HRTEM image and the *c-type* reflections occurring along [104]* of the FFTs (Figure 4d,e and Figure 5b). In fact, Christy and Larson [5] have demonstrated that, by considering twins in an underlying calcite matrix, even the doubled d_{012} corresponding to a 2·3.85 Å spacing can be generated. Thus, the doubled d_{104} and the d_{012} spacings measured on the thick part (>10 nm) of the HRTEM images can be used as evidence for nanosized calcite twins.

Nanotwins hosted in a matrix are commonly observed, in particular for materials such as diamond and SiC that contain abundant stacking faults [6,7,33,34]. In fact, the FFTs calculated from the thick region and the entire image show streaked reflections (white arrows in Figure 4d,e and Figure 5b), which can be associated with (10$\bar{1}$4) stacking faults that occur inside the 5–15 nm wide twin individuals.

3.3. Crystal Structure of the (10$\bar{1}$4) Calcite Twins

The crystallographic analysis suggests that the *c-type* reflections in calcite and their streaking can be successfully interpreted via (10$\bar{1}$4) twins and stacking faults. The structure model of this twin along [01$\bar{1}$0] (Figure 6a), built based on the work of Bruno et al. [35,36], reveals the orientation change of the CO_3 groups and the (10$\bar{1}$2) calcite planes across the twin interface, which agrees well with the observed features (Figures 4a and 5a). Based on geometry optimizations of 2D twinned slabs, Bruno et al. [35,36] demonstrated that the twinned individuals are translated and, at the twin interface, tilting and rotation of the CO_3 groups occur, which gives rise to variations of the Ca−O distances. Following the construction of the twin model, the crystal structure of a (10$\bar{1}$4) stacking fault can also be generated (Figure 6b).

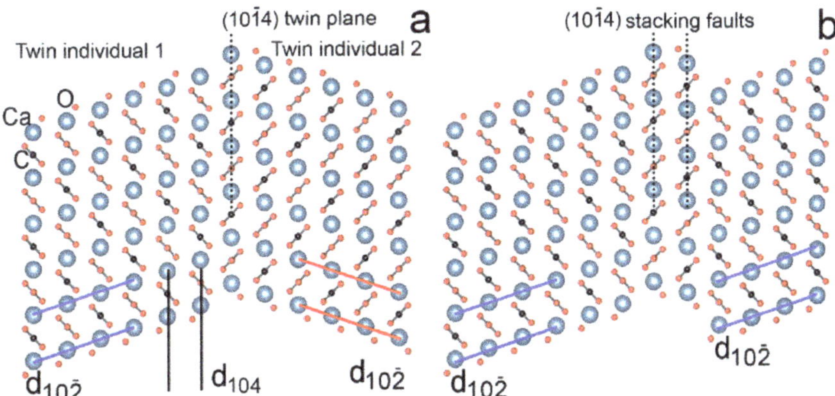

Figure 6. (a) Structure model of (10$\bar{1}$4) twin (**a**) and (10$\bar{1}$4) stacking faults (**b**) constructed based on the work of Bruno et al. [35,36]. In order, blue, red and black dots represent Ca, O and C atoms.

3.4. Possible Origin of the (10$\bar{1}$4) Calcite Twins

The (10$\bar{1}$4) twin is one of the four possible twins observed in calcite [37]. Although it has been reported as one of the main deformation twinning planes of calcite [37], it is also commonly associated with crystal growth and has been reported among others from the calcite shell of a sea urchin [5]. This twin was observed in a glendonite sample that formed from ikaite; thus, it may be associated with the ikaite-to-calcite transformation. However, since the formation of aragonite [17,18], amorphous calcium carbonate [19] and vaterite [20] has also been reported during this transition, a multiphase origin of the calcite twin is also plausible. In either way, interpretation based on the occurrence of this twin alone is definitely not a diagnostic for identifying glendonites and inferring cryogenic conditions [21].

4. Conclusions

c-type reflections (Figure 1) were identified in glendonite, a calcite pseudomorph after ikaite. As the sample was chemically pure $CaCO_3$ (Figure 2), it turned out that the *c-type* reflections cannot mimic the diffraction features associated with crystallographic ordering. TEM and crystallographic analysis suggested that, in fact, twinning by reticular pseudomerohedry (twin plane: $(10\bar{1}4)$; twin index: 4; obliquity: 0.74°) successfully explains these reflections (Figure 3). The twin individuals could be recognized from thin areas (~10 nm) of the HRTEM images through the direction change of the fringes corresponding to $d_{10\text{-}2}$ spacings (3.85Å). From the thicker part of the sample (>10 nm), the small-sized twin individual (5–10 nm) hosted in the calcite matrix and dynamically scattered electrons gave rise to double d_{104} spacings corresponding to $2 \cdot 3.03$ Å measured on HRTEM images (Figures 4 and 5). Streaked *c-type* reflections were also observed, and they were explained by $(10\bar{1}4)$ stacking faults occurring inside the small (5–15 nm wide) twin domains. It was shown that the orientation change of carbonate groups across the $(10\bar{1}4)$ calcite planes (Figure 6) gave rise to the $(10\bar{1}4)$ twins and stacking faults. Although in the glendonite sample Mg and Ca ordering could not occur, this phenomenon may be present in Mg containing calcite, from which *c-type* reflections have been reported. This paper strongly supports (1) the interpretation of Larsson and Christy [5] associating *c-type* reflections with $(10\bar{1}4)$ twins and with the orientation change of carbonate groups across the twin interface and (2) the final conclusion of considering the occurrence of $(10\bar{1}4)$ twins before attributing *c-type* reflections to Mg and Ca ordering.

Funding: The author is grateful for use of the TEM facility at the University of Pannonia and to its staff in the framework of grant no. GINOP-2.3.3-15-2016-0009 from the European Structural and Investments Funds. Financial support from the Hungarian National Research, Development and Innovation Office (projects ANN134433, FK123871 and 2019-2.1.11-TÉT-2019-00016), the Eötvös Loránd Research Network (NANOMIN projects, KEP-1/2020), as well as the János Bolyai Research Scholarship and the ÚNKP-20-5-PE-7 New National Excellence program of the Ministry for innovation and technology are acknowledged.

Data Availability Statement: Not applicable.

Acknowledgments: The author thanks Giovanni Ferraris for calculating the twinning features using the software GEMINOGRAPHY; his insightful comments and revision of this manuscript are also greatly appreciated. Paul Töchterle and Yuri Dublyansky are acknowledged for the glendonite sample. The author also thanks Zoltán May for measuring the glendonite sample with ICP-OES, István Dódony for emphasizing the orientation change of carbonate groups for the interpretation of HRTEM images, Dino Aquilano for discussing the $(10\bar{1}4)$ calcite twins and the useful comments of three anonymous reviewers.

Conflicts of Interest: The author declares no conflict of interest.

References

1. Inkson, B.J. Scanning electron microscopy (SEM) and transmission electron microscopy (TEM) for materials characterization. In *Materials Characterization Using Nondestructive Evaluation (NDE) Methods Editor(s): Gerhard Hübschen, Iris Altpeter, Ralf Tschuncky, Hans-Georg Herrmann*; Woodhead Publishing: Cambridge, UK, 2016; pp. 17–43. [CrossRef]
2. Jiang, Y.; Chen, Z.; Han, Y.; Deb, P.; Gao, H.; Xie, S.; Muller, D.A. Electron ptychography of 2D materials to deep sub-ångström resolution. *Nature* **2018**, *559*, 343–349. [CrossRef]
3. Yan, X.; Liu, C.; Gadre, C.A.; Gu, L.; Aoki, T.; Lovejoy, T.C.; Pan, X. Single-defect phonons imaged by electron microscopy. *Nature* **2021**, *589*, 65–69. [CrossRef]
4. Németh, P.; Dódony, I.; Pósfai, M.; Buseck, P.R. Complex defect in pyrite and its structure model derived from geometric phase analysis. *Microsc. Microanal.* **2013**, *19*, 1303–1307. [CrossRef] [PubMed]
5. Larsson, A.-K.; Christy, A.G. On twinning and microstructures in calcite and dolomite. *Am. Mineral.* **2008**, *93*, 103–113. [CrossRef]
6. Németh, P.; Garvie, L.A.J.; Buseck, P.R. Twinning of cubic diamond explains reported nanodiamond polymorphs. *Sci. Rep.* **2015**, *5*, 18381. [CrossRef] [PubMed]
7. Cayron, C. Diffraction artefacts from twins and stacking faults, and the mirage of hexagonal, polytypes or other superstructures. *Scr. Mater.* **2020**, *194*, 113629. [CrossRef]

8. Huggett, J.M.; Schultz, B.P.; Shearman, D.J.; Smith, A.J. The petrology of ikaite pseudomorphs and their diagenesis. *Proc. Geol. Assoc.* **2005**, *116*, 207–220. [CrossRef]

9. De Lurio, J.L.; Frakes, L.A. Glendonites as a paleoenvironmental tool: Implications for early Cretaceous high latitude climates in Australia. *Geochim. Cosmochim. Acta* **1999**, *633*, 1039–1048. [CrossRef]

10. Rogov, M.; Ershova, V.; Vereshchagin, O.; Vasileva, K.; Mikhailova, K.; Krylov, A. Database of global glendonite and ikaite records throughout the Phanerozoic. *Earth Syst. Sci. Data* **2021**, *13*, 343–356. [CrossRef]

11. Tollefsen, E.; Balic-Zunic, T.; Mörth, C.M.; Brüchter, V.; Lee, C.C.; Skelton, A. Ikaite nucleation at 35 °C challenges the use of glendonite as a paleotemperature indicator. *Sci. Rep.* **2000**, *10*, 8141. [CrossRef]

12. Brooks, R.; Clark, L.M.; Thurston, E.F. Calcium carbonate and its hydrates. *Philos. Trans. R. Soc. A* **1950**, *243*, 145–167. [CrossRef]

13. Swainson, I.P.; Hammond, R.P. Ikaite, CaCO3.6H2O: Cold comfort for glendonites as palaeothermometers. *Am. Mineral.* **2001**, *86*, 1530–1533. [CrossRef]

14. Selleck, B.W.; Carr, P.F.; Jones, B.G. A Review and Synthesis of Glendonites (Pseudomorphs after Ikaite) with New Data: Assessing Applicability as Recorders of Ancient Coldwater Conditions. *J. Sediment. Res.* **2007**, *77*, 980–991. [CrossRef]

15. Pauly, H. 'Ikaite', a new mineral from Greenland. *Arctic Res.* **1963**, *16*, 263–279. [CrossRef]

16. Bischoff, J.L.; Fitzpatrick, J.A.; Rosenbauer, R.J. The solubility and stabilization of ikaite (CaCO3· 6H2O) from 0 to 25 C: Environmental and paleoclimatic implications for thinolite tufa. *J. Geol.* **1993**, *101*, 21–33. [CrossRef]

17. Stein, E.S.; Smith, A.J. Authigenic carbonate nodules in the Nankai Trough, site 583. *Initial. Rep. Deep. Sea Drill. Proj.* **1985**, *87*, 659–668.

18. Council, T.C.; Bennett, P.C. Geochemistry of ikaite formation at Mono Lake, California: Implications for the origin of tufa mounds. *Geology* **1993**, *21*, 971–974. [CrossRef]

19. Zou, Z.; Bertinetti, L.; Habraken, W.; Fratzl, P. Reentrant phase transformation from crystalline ikaite to amorphous calcium carbonate. *CrystEngComm* **2018**, *20*, 2902–2906. [CrossRef]

20. Tang, C.C.; Thompson, S.P.; Parker, J.E.; Lennie, A.R.; Azough, F.; Kato, K. The ikaite-to-vaterite transformation: New evidence from diffraction and imaging. *J. Appl. Crystallogr.* **2009**, *42*, 225–233. [CrossRef]

21. Németh, P.; Töchterle, P.; Dublyansky, Y.; Stalder, R.; Molnár, Z.S.; Klébert, S.Z. Spötl Tracing structural relicts of the ikaite-to-calcite transformation in cryogenic cave glendonite. *Am. Mineral* **2021**. submitted.

22. Reeder, R.J.; Wenk, H.R. Microstructures in low-temperature dolomites. *Geophys. Res. Lett.* **1979**, *6*, 77–80. [CrossRef]

23. Redfern, S.A.T.; Salje, E.; Navrotsky, A. High-temperature enthalpy at the orientational order-disorder transition in calcite: Implications for the calcite/aragonite phase equilibrium. *Contrib. Mineral. Petrol.* **1989**, *101*, 479–484. [CrossRef]

24. Van Tendeloo, G.; Wenk, H.R.; Gronsky, R. Modulated structures in calcium dolomite: A study by electron microscopy. *Phys. Chem. Miner.* **1985**, *12*, 333–341. [CrossRef]

25. Wenk, H.R.; Meisheng, H.; Lindsey, T.; Morris, J.W. Superstructure in ankerite and calcite. *Phys. Chem. Miner.* **1991**, *17*, 527–539. [CrossRef]

26. Shen, Z.; Konishi, H.; Brown, P.E.; Xu, H. STEM investigation of exsolution lamellae and "c" reflections in Ca-rich dolomite from the Platteville Formation, western Wisconsin. *Am. Mineral.* **2013**, *4*, 760–766. [CrossRef]

27. Nespolo, M.; Ferraris, G. Applied geminography—Symmetry analysis of twinned crystals and definition of twinning by reticular polyholohedry. *Acta Crystallogr. A* **2004**, *60*, 89–95. [CrossRef]

28. Dublyansky, Y.; Moseley, G.E.; Lyakhnitsky, Y.; Cheng, H.; Edwards, R.L.; Scholz, D.; Spötl, C. Late Palaeolithic cave art and permafrost in the Southern Ural. *Sci. Rep.* **2018**, *8*, 12080. [CrossRef] [PubMed]

29. Seto, Y.D.; Hamane, T.N.; Sata, N. Development of a software suite on X-ray diffraction experiments. *Rev. High Press. Sci. Technol.* **2010**, *20*, 269–276. [CrossRef]

30. Zsigmond, A.R.; Kántor, I.; May, Z.; Urák, I.; Héberger, K. Elemental composition of Russula cyanoxantha along an urbanization gradient in Cluj-Napoca (Romania). *Chemosphere* **2020**, *238*, 124566. [CrossRef]

31. Momma, K.; Izumi, F. VESTA 3 for three-dimensional visualization of crystal, volumetric and morphology data. *J. Appl. Crystallogr.* **2011**, *44*, 1272–1276. [CrossRef]

32. Nespolo, M.; Ferraris, G. The derivation of twin laws in non-merohedric twins—Application to the analysis of hybrid twins. *Acta Crystallogr. A* **2006**, *62*, 336–349. [CrossRef]

33. Németh, P.; Garvie, L.A.J.; Aoki, T.; Dubrovinskaia, N.; Dubrovinsky, L.; Buseck, P.R. Lonsdaleite is faulted and twinned cubic diamond and does not exist as a discrete material. *Nat. Commun.* **2014**, *5*, 5447. [CrossRef] [PubMed]

34. Daulton, T.L.; Bernatowicz, T.J.; Lewis, R.S.; Messenger, S.; Stadermann, F.J.; Amari, S. Polytype distribution in circumstellar silicon carbide: Microstructural characterization by transmission electron microscopy. *Geochim. Cosmochim. Acta* **2003**, *67*, 4743–4767. [CrossRef]

35. Bruno, M.; Massaro, F.R.; Rubbo, M.; Prencipe, M.; Aquilano, D. The (10.4), (01.8), (01.2) and (00.1) twin laws of calcite (CaCO3): Equilibrium geometry of the twin boundary interfaces and twinning energy. *Cryst. Growth Des.* **2010**, *10*, 3102–3109. [CrossRef]

36. Aquilano, D.; Benages-Vilau, R.; Bruno, M.; Rubbo, M.; Massaro, F. Positive {*hk.l*} and negative {*hk.l*} forms of calcite (CaCO3) crystal. New open questions from the evaluation of their surface energies. *CrystEngComm* **2013**, *15*, 4465. [CrossRef]

37. Wenk, H.R.; Barber, D.J.; Reeder, R.J. Microstructures in Carbonates. In *Carbonates: Mineralogy and Chemistry*; Reeder, R.J., Ed.; Reviews in Mineralogy; Mineralogical Society of America: Blacksburg, VA, USA, 1983; Volume 11, pp. 301–367. [CrossRef]

Article

Twinning, Superstructure and Chemical Ordering in Spryite, $Ag_8(As^{3+}_{0.50}As^{5+}_{0.50})S_6$, at Ultra-Low Temperature: An X-Ray Single-Crystal Study

Luca Bindi [1,*] and Marta Morana [2]

1   Dipartimento di Scienze della Terra, Università degli Studi di Firenze, Via G. La Pira 4, I-50121 Firenze, Italy
2   Department of Chemistry, University of Pavia, via Taramelli 12, 27100 Pavia, Italy; marta.morana@unipv.it
*   Correspondence: luca.bindi@unifi.it

Abstract: Spryite $(Ag_{7.98}Cu_{0.05})_{\Sigma=8.03}(As^{5+}_{0.31}Ge_{0.36}As^{3+}_{0.31}Fe^{3+}_{0.02})_{\Sigma=1.00}S_{5.97}$, and ideally $Ag_8(As^{3+}_{0.5}As^{5+}_{0.5})S_6$, is a new mineral recently described from the Uchucchacua polymetallic deposit, Oyon district, Catajambo, Lima Department, Peru. Its room temperature structure exhibits an orthorhombic symmetry, space group $Pna2_1$, with lattice parameters $a = 14.984(4)$, $b = 7.474(1)$, $c = 10.571(2)$ Å, $V = 1083.9(4)$ Å^3, $Z = 4$, and shows the coexistence of As^{3+} and As^{5+} distributed in a disordered fashion in a unique mixed position. To analyze the crystal-chemical behaviour of the arsenic distribution at ultra-low temperatures, a structural study was carried out at 30 K by means of in situ single-crystal X-ray diffraction data (helium-cryostat) on the same sample previously characterized from a chemical and structural point of view. At 30 K, spryite still crystallizes with orthorhombic symmetry, space group $Pna2_1$, but gives rise to a a $\times$ 3b $\times$ c superstructure, with $a = 14.866(2)$, $b = 22.240(4)$, $c = 10.394(1)$ Å, $V = 3436.5(8)$ Å^3 and $Z = 4$ ($Ag_{24}As^{3+}As^{5+}Ge^{4+}S_{18}$ stoichiometry). The twin laws making the twin lattice simulating a perfect hexagonal symmetry have been taken into account and the crystal structure has been solved and refined. The refinement of the structure leads to a residual factor $R = 0.0329$ for 4070 independent observed reflections [with $F_o > 4\sigma(F_o)$] and 408 variables. The threefold superstructure arises from the ordering of As^{3+} and (As^{5+}, Ge^{4+}) in different crystal-chemical environments.

Keywords: twinning; chemical ordering; superstructure; spryite; argyrodite-type compounds; ultra-low temperature

Citation: Bindi, L.; Morana, M. Twinning, Superstructure and Chemical Ordering in Spryite, $Ag_8(As^{3+}_{0.50}As^{5+}_{0.50})S_6$, at Ultra-Low Temperature: An X-Ray Single-Crystal Study. *Minerals* **2021**, *11*, 286. https://doi.org/10.3390/min11030286

Academic Editor: Giovanni Ferraris

Received: 3 February 2021
Accepted: 8 March 2021
Published: 10 March 2021

Publisher's Note: MDPI stays neutral with regard to jurisdictional claims in published maps and institutional affiliations.

1. Introduction

Spryite, ideally $Ag_8(As^{3+}_{0.5}As^{5+}_{0.5})S_6$, is a new mineral belonging to the argyrodite group recently described from the Uchucchacua polymetallic deposit, Oyon district, Catajambo, Lima Department, Peru [1]. Argyrodite-type compounds allow for a large structural and chemical heterogeneity with the general formula $A^{m+}_{[(12-n-y)/m]}B^{n+}Q^{2-}_{6-y}X^-_{y}$, where A is a mono- or di-valent twofold or threefold coordinated cation such as Cu^+, Ag^+, Li^+, Cd^{2+}, Hg^{2+}, B is a tri-, tetra-, or penta-valent tetrahedral cation, like Al^{3+}, Ga^{3+}, Si^{4+}, Ge^{4+}, Sn^{4+}, Ti^{4+}, P^{5+}, As^{5+}, Sb^{5+}, Nb^{5+}, Ta^{5+}, and Q and X are respectively chalcogenide or halide anions [2]. Natural and synthetic argyrodites have drawn attention over time for their peculiar physical properties, such as superionic conduction, and their potential application as electrolytes, e.g., [3]. In this respect, spryite represents an exception since it does not behave as a superionic conductor [1]. The absence of such property was attributed to the presence of As^{3+} in the structure, another unique feature of spryite. As^{3+} was not considered as a possible cation for the B site in the argyrodite-type structure because of the presence of the lone-pair electrons that do not allow tetrahedral coordination. It is thus not surprising that, before the description of spryite, only Al^{3+}, Ga^{3+}, and Fe^{3+} were reported as trivalent cations in argyrodites [4–6]. In spryite trivalent and pentavalent, As (together with Ge^{4+}) share the same split sites with slightly different atomic coordinates. On the one hand, As^{3+} is split toward three S atoms to produce AsS_3 pyramids, analogously to sulfosalts [7]. On the other hand, As^{5+} and Ge^{4+} sit in the tetrahedral site typically occupied by Ge in argyrodite-type compounds [8]. The presence of disordered $As^{3+}S_3$ pyramid might be responsible for the absence of superionic conduction, since they can hinder the mobility of the Ag cations that in argyrodites-like structures are highly delocalized over all the available sites, even at room temperature [9]. In fact, this mineral is characterized by a network of non-interacting Ag cation, with all sites fully occupied. Temperature dependence is another peculiar aspect of spryite. Argyrodite-type compounds usually show three phase transitions as a function of temperature. The high temperature phase has space group *F-43m*, that transforms to $P2_13$. On further cooling, these compounds apparently adopt again the *F-43m* space group, but actually have an orthorhombic symmetry, such as $Pna2_1$, *Pnam*, or $Pmn2_1$ [7]. Conversely, spryite was shown to maintain the orthorhombic structure from 90 to 500 K [1], thus representing a unique case in the argyrodite family. Considering the peculiar structural features and temperature dependence of spryite, we investigated its crystal structure at 30 K, in order to understand the effects of ultra-low temperature on the structure and to verify if the disordered As/Ge position present in the room-temperature structure could give rise to some localized ordering and thus to a possible superstructure at ultra-low temperature.

2. X-ray Crystallography

The same crystal of spryite used to study the temperature behaviour in the range 90–500 K (chemical composition $(Ag_{7.98}Cu_{0.05})_{\Sigma=8.03}(As^{5+}_{0.31}Ge_{0.36}As^{3+}_{0.31}Fe^{3+}_{0.02})_{\Sigma=1.00}S_{5.97}$ [1]) was mounted on an Oxford Diffraction Xcalibur 3 diffractometer (Oxford Diffraction, Oxford, UK) (Enhance X-ray source, X-ray $MoK\alpha$ radiation, λ = 0.71073 Å), fitted with a Sapphire 2 CCD detector and an Oxford cryostream cooler (helium-cryostat). The temperature was lowered at 30 K and, before the measurement, the sample was held at that temperature for about 30 min. The diffraction pattern at 30 K was consistent with an orthorhombic symmetry but additional reflections leading to a threefold $\mathbf{a}\times3\mathbf{b}\times\mathbf{c}$ commensurate superstructure were observed (i.e., $a \approx 14.9$ Å, $b \approx 22.2$, $c \approx 10.4$ Å). To account for a potential reduction of symmetry for the low-temperature structure of spryite and given the fact that the crystal is intimately twinned, a relatively high $\sin(\theta)/\lambda$ cutoff and a high redundancy were chosen in the recording setting design. Intensity integration and standard Lorentz-polarization correction were performed with the *CrysAlis* software package [10,11]. The diffraction quality was found to be excellent, thus indicating that no deterioration of the crystal occurred even at ultra-low temperature. All the collected data are plotted down the c-axis and shown in Figure 1. A strong hexagonal pseudo-symmetry of the X-ray reflections is evident, which is due to the pervasive twinning giving rise to a pseudo-cubic, face-centered cell with $a \approx 10.5$ Å at room temperature [1] and to a pseudo-hexagonal cell with $a \approx 7.5$ and $c \approx 10.5$ Å at 30 K.

The values of the equivalent pairs were averaged. The merging R for the ψ-scan data set decreased from 0.1506 before absorption correction to 0.0355 after this correction. The analysis of the systematic absences (*0kl*: $k + l = 2n$; *h0l*: $h = 2n$; *h00*: $h = 2n$; *0k0*: $k = 2n$; *00l*: $l = 2n$) are consistent with the space groups *Pnam* (*Pnma* as standard) and $Pna2_1$. Statistical tests on the distribution of $|E|$ values strongly indicate the absence of an inversion centre $[\,|E^2-1|$ = 0.695], thus suggesting the choice of the space group $Pna2_1$. To decide the correct space group for the low-temperature superstructure we also analyzed the maximum orthorhombic *klassengleiche* subgroups of the $Pna2_1$ room-temperature space group. We noticed that there is only one subgroup with $\mathbf{b'} = 3\mathbf{b}$, that is $Pna2_1$, and thus the superstructure was solved in this space group.

The position of most of the atoms was determined from the three-dimensional Patterson synthesis. A least-squares refinement, by means of the program SHELXL-97 [12], using these heavy-atom positions and isotropic temperature factors, yielded an R factor of 0.2005. Three-dimensional difference Fourier synthesis yielded the position of the remaining sulfur atoms. The introduction of anisotropic temperature factors for all the atoms led to $R = 0.0329$ for 4070 observed reflections $[F_o > 4\sigma(F_o)]$ and $R = 0.0346$ for all 4705 independent reflections. Neutral scattering factors for Ag, As, Ge, and S were taken from the International Tables for X-ray Crystallography [13].

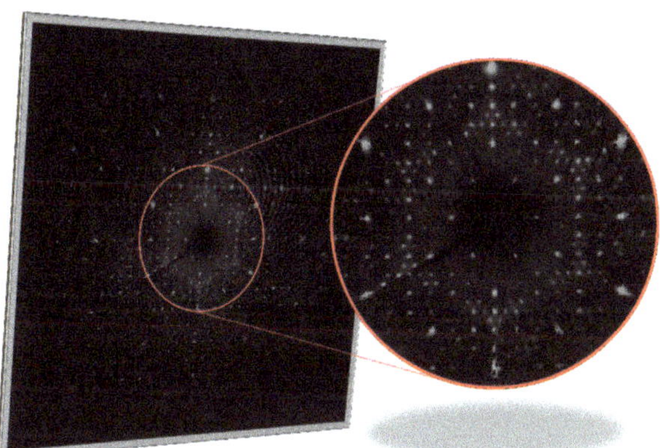

Figure 1. Hexagonal pseudo-symmetry in the X-ray diffraction pattern of spryite at 30 K due to the pervasive twinning.

In order to better model the twinning occurring in spryite [1], we then took into account the twin law, which makes the twin lattice (L_T) hexagonal (twinning by reticular merohedry [14]) using the program JANA2006 [15]. For details of this kind of twinning and on the averaging of equivalent reflections for twins in JANA, see the appendix in [16]. Remarkably, the same set of twin matrices (referred to the orthorhombic cell) were used either at 30 K or at room temperature [1]. Once again, the structure refinement was initiated in the orthorhombic supercell. After several cycles, the structure could be smoothly refined without any damping factor or restrictions. The residual value converged to $R = 0.0329$ for 4070 observed reflections $[F_o > 4\sigma(F_o)]$ and $R = 0.0346$ including all the 4705 collected reflections in the refinement. Inspection of the difference Fourier map revealed that maximum positive and negative peaks were 1.90 a-d 1.67 e$^-$/Å^3, respectively. Experimental details and R indices are given in Table 1. Fractional atomic coordinates and isotropic displacement parameters are shown in Table 2. The Crystallographic Information File (CIF)of the structure is deposited as Supplementary Materials.

Table 1. Experimental and refinement details for spryite at 30 K.

Spryite	
Temperature	30(2) K
Wavelength	0.71073 Å
Crystal system	Orthorhombic
Space group	$Pna2_1$
Unit cell dimensions	$a = 14.866(2)$ Å
	$b = 22.240(4)$ Å
	$c = 10.3940(10)$ Å
Volume	3436.5(8) Å^3
Z	4
Density (calculated)	6.549 Mg/mm^3
Crystal size	$0.040 \times 0.030 \times 0.020$ mm^3
Theta range for data collection	4.78 to 30.00°
h,k,l ranges	$-20 \leq h \leq 20, -31 \leq k \leq 31, -14 \leq l \leq 14$
Reflections collected	38705
Independent reflections	4705 [R_{int} = 0.0355]
Data/restraints/parameters	4705/1/408
Goodness of fit on F^2	1.052
Final R indices [$F_o > 4\sigma(F_o)$]	$R1$ = 0.0329, w$R2$ = 0.0670
R indices (all data)	$R1$ = 0.0346, w$R2$ = 0.0678
Extinction coefficient	0.000301(17)
Largest diff. peak and hole	1.90 and -1.67 e.Å^{-3}
Twin matrices (referred to the orthorhombic cell)	$\begin{bmatrix} 1 & 0 & 0 \\ 0 & 1 & 0 \\ 0 & 0 & 1 \end{bmatrix} \begin{bmatrix} 1/2 & 1/4 & 1/2 \\ -1 & -1/2 & 1 \\ 1 & -1/2 & 0 \end{bmatrix} \begin{bmatrix} 1/2 & -1/4 & 1/2 \\ 1 & -1/2 & -1 \\ 1 & 1/2 & 0 \end{bmatrix}$ $\begin{bmatrix} 0 & -1/2 & 0 \\ 2 & 0 & 0 \\ 0 & 0 & -1 \end{bmatrix} \begin{bmatrix} -1/2 & 1/4 & -1/2 \\ 1 & -1/2 & -1 \\ 1 & 1/2 & 0 \end{bmatrix} \begin{bmatrix} 1/2 & 1/4 & -1/2 \\ 1 & 1/2 & 1 \\ -1 & 1/2 & 0 \end{bmatrix}$
Twin fractions	0.28(3), 0.20(2), 0.16(2), 0.14(2), 0.12(2), 0.10(2)

Table 2. Atom coordinates and equivalent isotropic displacement parameters (Å^2) for spryite.

Site	x/a	y/b	z/c	U_{eq}
Ag1A	0.1257(3)	0.07361(17)	0.3742(5)	0.0383(5)
Ag1B	0.1269(2)	0.40562(13)	0.3764(4)	0.0359(4)
Ag1C	0.1269(4)	0.7341(3)	0.3739(6)	0.0393(6)
Ag2A	0.06183(4)	0.07485(2)	0.83671(6)	0.02599(11)
Ag2B	0.05993(4)	0.40511(2)	0.83817(6)	0.02494(11)
Ag2C	0.06321(3)	0.73552(2)	0.83630(6)	0.02402(11)
Ag3A	0.43197(5)	0.02050(3)	0.01921(7)	0.03304(14)
Ag3B	0.43246(4)	0.35041(3)	0.01990(7)	0.02850(13)
Ag3C	0.43262(5)	0.68067(4)	0.01956(9)	0.03866(17)
Ag4A	0.2768(4)	0.1662(3)	0.0834(8)	0.0290(10)
Ag4B	0.27687(4)	0.50675(14)	0.08245(6)	0.0286(4)
Ag4C	0.2770(5)	0.8364(4)	0.0819(9)	0.0391(16)
Ag5A	0.41845(5)	0.03021(3)	0.69619(9)	0.03584(14)
Ag5B	0.41848(4)	0.36017(3)	0.69717(8)	0.03396(13)
Ag5C	0.41868(4)	0.70034(3)	0.69693(7)	0.02805(12)
Ag6A	0.27281(4)	0.12823(3)	0.68399(7)	0.02865(12)
Ag6B	0.27274(5)	0.46785(3)	0.68435(9)	0.03504(13)
Ag6C	0.27304(4)	0.79812(3)	0.68428(7)	0.02883(12)

Table 2. *Cont.*

Site	x/a	y/b	z/c	U_{eq}
Ag7A	0.01698(4)	0.0045(6)	0.60282(8)	0.0387(10)
Ag7B	0.0175(4)	0.3339(3)	0.6021(6)	0.0249(9)
Ag7C	0.0168(5)	0.6639(3)	0.6036(8)	0.0351(15)
Ag8A	0.25877(4)	0.04283(3)	0.90577(8)	0.03030(13)
Ag8B	0.25881(4)	0.37291(3)	0.90561(8)	0.03138(13)
Ag8C	0.25860(4)	0.70268(3)	0.90549(7)	0.02573(12)
As^{5+}	0.37565(4)	0.07852(3)	0.34985(10)	0.02252(13)
Ge	0.37583(4)	0.40121(3)	0.34959(10)	0.01968(12)
As^{3+}	0.37990(3)	0.78942(3)	0.29965(11)	0.02286(12)
S1A	0.1227(8)	0.1644(5)	0.9763(8)	0.030(3)
S1B	0.12244(14)	0.5046(18)	0.9770(3)	0.037(3)
S1C	0.1221(10)	0.8341(6)	0.9774(13)	0.046(4)
S2A	−0.00394(18)	0.09037(11)	0.2315(3)	0.0388(5)
S2B	−0.00418(18)	0.42044(11)	0.2316(3)	0.0380(5)
S2C	−0.00434(16)	0.75029(10)	0.2313(3)	0.0358(5)
S3A	0.37300(17)	0.15601(12)	0.4794(3)	0.0362(5)
S3B	0.37279(16)	0.48499(10)	0.4791(3)	0.0379(5)
S3C	0.37272(16)	0.81480(14)	0.4788(3)	0.0327(5)
S4A	0.25826(17)	0.07751(13)	0.2306(3)	0.0380(5)
S4B	0.25859(18)	0.40740(13)	0.2300(3)	0.0402(5)
S4C	0.25859(18)	0.73719(13)	0.2306(3)	0.0407(5)
S5A	0.3861(7)	0.1054(4)	0.8663(12)	0.0334(10)
S5B	0.3867(9)	0.4357(5)	0.8691(16)	0.0322(11)
S5C	0.3855(13)	0.7631(10)	0.879(2)	0.0398(19)
S6A	0.12309(13)	0.08887(10)	0.6117(3)	0.0320(4)
S6B	0.12280(14)	0.42869(10)	0.6117(3)	0.0351(5)
S6C	0.12278(14)	0.75866(10)	0.6112(3)	0.0337(4)

3. Description of the Low-Temperature Structure and Discussion

The low-temperature structure solution of spryite showed that the atomic arrangement of the mineral at 30 K is topologically identical to that observed at room temperature [1] with the cation ordering being limited to small portions of the structure only. Indeed, the solution revealed that As^{3+}, As^{5+}, and Ge^{4+} are ordered into three specific sites. Indeed, the unique mixed (As, Ge) position in the room-temperature structure (Wyckoff position 4*a*) transforms into three 4*a* Wyckoff positions in the low-temperature structure hosting As^{3+}, As^{5+}, and Ge, respectively. This does not lead to any reduction of symmetry as the space group ($Pna2_1$) remains the same as the room-temperature structure. In the structure, Ag occupies sites with coordination ranging from quasi-linear to almost tetrahedral connected into a framework (Figure 2 and Table 3). In particular, 10 Ag atoms are fourfold coordinated, 11 are threefold and 3 are twofold coordinated. As in the ambient temperature structure, the average bond length increases from the twofold to the fourfold coordination: the average Ag-S distance ranges from 2.569 to 2.722 Å for the almost tetrahedral geometry, from 2.475 to 2.562 Å for the trigonal geometry, and from 2.337 to 2.428 Å for the quasi-linear geometry. Each Ag site gives rise to three sites in the ultra-low temperature structure maintaining the coordination present at room temperature, e.g., the fourfold coordinated Ag2 corresponds to three fourfold coordinated Ag2A, Ag2C, Ag3C. The almost tetrahedral Ag1 is the only exception, since it is related to a fourfold coordinated Ag1C and two threefold coordinated Ag1A and Ag1B. The analysis of the crystal-chemical characteristic of the Ag-environments indicates that the As/Ge chemical ordering observed in the low-temperature crystal structure of spryite does not affect significantly the geometry of their coordination polyhedra, highlighting a clear similarity with the low temperature structure.

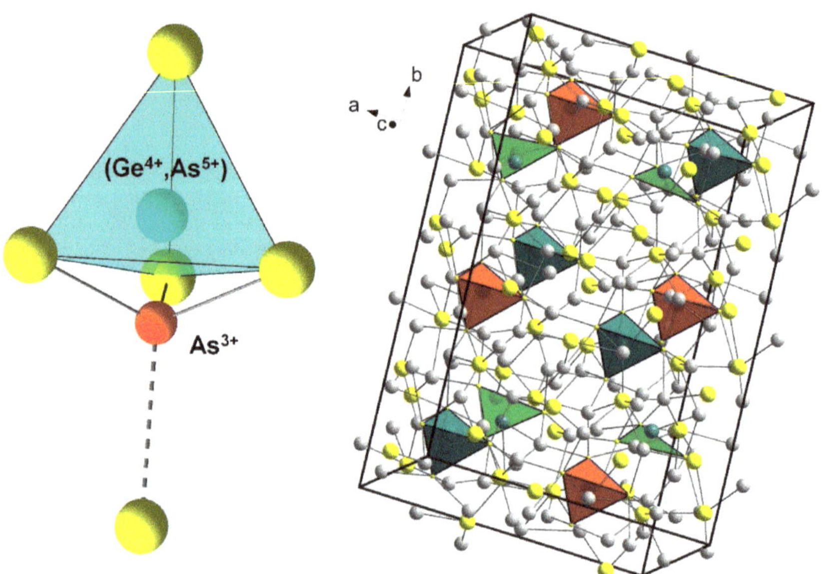

Figure 2. Left: Portion of the crystal structure of spryite at room temperature highlighting the disorder between the *M*1 and *M*2 positions (As^{3+}, As^{5+} and Ge^{4+}) at room temperature. In light blue the *M*1 tetrahedron (As^{5+} and Ge^{4+}) and with a red sphere the *M*2 atom (As^{3+}). **Right:** The crystal structure of spryite at 30 K; Ag and S atoms are given as white and yellow spheres, respectively. Light blue and red tetrahedra are filled by As^{5+} and Ge^{4+}, respectively, whereas green polyhedra are filled by As^{3+}. The unit cell and the orientation of the structure are outlined.

Table 3. Selected bond distances [Å] and angles (°) for spryite at 30K.

Atoms	Distance	Atoms	Distance
Ag1A-S2A	2.460(6)	Ag6A-S6A	2.507(2)
Ag1A-S4A	2.474(5)	Ag6A-S4C	2.515(3)
Ag1A-S6A	2.492(6)	Ag6A-S5A	2.586(12)
<Ag1A-S>	**2.475**	Ag6A-S3A	2.669(3)
Ag1B-S4B	2.480(5)	**<Ag6A-S>**	**2.569**
Ag1B-S2B	2.484(4)	Ag6B-S6B	2.510(2)
Ag1B-S6B	2.499(5)	Ag6B-S4A	2.528(3)
<Ag1B-S>	**2.488**	Ag6B-S3B	2.629(3)
Ag1C-S4C	2.461(6)	Ag6B-S5B	2.659(16)
Ag1C-S2C	2.477(7)	**<Ag6B-S>**	**2.582**
Ag1C-S6C	2.527(6)	Ag6C-S6C	2.517(2)
Ag1C-S5A	2.870(12)	Ag6C-S4B	2.521(3)
<Ag1C-S>	**2.584**	Ag6C-S3C	2.626(3)
Ag2A-S6A	2.529(3)	Ag6C-S5C	2.74(2)
Ag2A-S1A	2.624(11)	**<Ag6C-S>**	**2.601**
Ag2A-S5B	2.636(13)	Ag7A-S6A	2.453(10)
Ag2A-S3B	2.670(3)	Ag7A-S2A	2.506(12)
<Ag2A-S>	**2.615**	Ag7A-S3B	2.511(3)
Ag2B-S6B	2.587(3)	**<Ag7A-S>**	**2.490**
Ag2B-S5A	2.611(11)	Ag7B-S2C	2.313(6)
Ag2B-S3C	2.678(3)	Ag7B-S3A	2.509(6)
Ag2B-S1B	2.80(3)	Ag7B-S6B	2.627(6)
<Ag2B-S>	**2.699**	**<Ag7B-S>**	**2.483**
Ag2C-S3A	2.497(3)	Ag7C-S2B	2.307(8)

Table 3. *Cont.*

Atoms	Distance	Atoms	Distance
Ag2C-S6C	2.554(3)	Ag7C-S3C	2.548(8)
Ag2C-S5C	2.68(2)	Ag7C-S6C	2.634(7)
Ag2C-S1C	2.780(13)	**<Ag7C-S>**	**2.496**
<Ag2C-S>	**2.628**	Ag8A-S5A	2.384(7)
Ag3A-S6B	2.399(3)	Ag8A-S3B	2.462(3)
Ag3A-S5A	2.560(11)	**<Ag8A-S>**	**2.428**
Ag3A-S2B	2.739(3)	Ag8B-S5B	2.389(10)
Ag3A-S1B	2.919(8)	Ag8B-S3C	2.464(3)
<Ag3A-S>	**2.654**	**<Ag8B-S>**	**2.427**
Ag3B-S6C	2.396(3)	Ag8C-S5C	2.33(2)
Ag3B-S5B	2.553(14)	Ag8C-S3A	2.344(3)
Ag3B-S2A	2.732(3)	**<Ag8c-S>**	**2.337**
Ag3B-S1A	2.882(12)	As^{5+}-S1B	2.11(3)
<Ag3B-S>	**2.722**	As^{5+}-S4A	2.141(3)
Ag3C-S6A	2.402(3)	As^{5+}-S2B	2.168(3)
Ag3C-S5C	2.45(2)	As^{5+}-S3A	2.187(3)
Ag3C-S2C	2.843(3)	**<As^{5+}-S>**	**2.152**
Ag3C-S1C	2.870(15)	S1B-As^{5+}-S4A	111.4(4)
<Ag3C-S>	**2.641**	S1B-As^{5+}-S2B	110.6(3)
Ag4A-S4A	2.511(8)	S4A-As^{5+}-S2B	110.09(12)
Ag4A-S1A	2.548(14)	S1B-As^{5+}-S3A	103.2(7)
Ag4A-S6C	2.557(7)	S4A-As^{5+}-S3A	110.48(11)
<Ag4A-S>	**2.539**	S2B-As^{5+}-S3A	110.81(10)
Ag4B-S6A	2.375(3)	Ge-S1C	1.998(13)
Ag4B-S1B	2.544(2)	Ge-S4B	2.145(3)
Ag4B-S4B	2.703(4)	Ge-S2A	2.177(3)
<Ag4B-S>	**2.541**	Ge-S3B	2.299(3)
Ag4C-S1C	2.545(18)	**<Ge-S>**	**2.155**
Ag4C-S6B	2.556(8)	S1C-Ge-S4B	116.5(4)
Ag4C-S4C	2.707(10)	S1C-Ge-S2A	115.3(5)
<Ag4C-S>	**2.551**	S4B-Ge-S2A	109.55(13)
Ag5A-S1B	2.426(10)	S1C-Ge-S3B	102.5(4)
Ag5A-S5A	2.481(12)	S4B-Ge-S3B	105.74(10)
Ag5A-S2B	2.778(3)	S2A-Ge-S3B	106.06(9)
<Ag5A-S>	**2.562**	As^{3+}-S3C	1.949(3)
Ag5B-S1C	2.433(13)	As^{3+}-S2C	2.061(2)
Ag5B-S5B	2.498(15)	As^{3+}-S4C	2.262(3)
Ag5B-S2C	2.780(2)	As^{3+}-S3C	1.949(3)
<Ag5B-S>	**2.570**	**<As(3)-S>**	**2.055**
Ag5C-S5C	2.40(2)		
Ag5C-S1A	2.505(9)		
Ag5C-S2A	2.778(3)		
<Ag5C-S>	**2.561**		

Bold = mean poyhedral bond distances.

As^{3+} forms AsS_3 pyramids, typical of sulfosalts, and (Ge^{4+}, As^{5+}) links four sulfur atoms in a tetrahedral coordination. Given the close scattering power between As and Ge—and their corresponding site geometry—it is hard to say which tetrahedron is dominated by As (or Ge). Bond-valence considerations do not help in this case as the mean tetrahedral distances and the mean tetrahedral angles are very close: 2.152 and 2.155 Å and 109.4 and 110.9°, for the AsS_4 and GeS_4, respectively (Table 3). However, the slightly smaller value for the AsS_4 tetrahedron seems in agreement with the slightly shorter As^{5+}-S distance (2.169 Å—see discussion in [17]) than that observed for GeS_4 in pure argyrodite (i.e., 2.212 Å; [8]). Furthermore, the analysis of the angle variance (σ^2) and the quadratic elongation (λ) of the two tetrahedra [18] reveals strong differences: AsS_4 tetrahedra exhibit a $\sigma^2 = 9.86$ and $\lambda = 1.0030$, while GeS_4 tetrahedra exhibit a $\sigma^2 = 31.33$ and $\lambda = 1.0094$. The general higher distortion introduced by the entry of Ge^{4+} in crystal structures [8] represents a further corroboration of the right assignment of As^{5+} and Ge^{4+} in the two tetrahedra of spryite at ultra-low temperature.

4. Conclusions

Our investigation shows that the crystal structure of cooled spryite is very close to that observed at room temperature. Spryite is also characterized by pervasive twinning, thus requiring an accurate structural characterization. We demonstrate by means of an in situ data collection at 30 K that there is an ordering between As^{3+} and (As^{5+}, Ge^{4+}), leading to a threefold superstructure. Spryite confirms its uniqueness in the argyrodite family of compounds, since it maintains its orthorhombic symmetry on a large temperature range, together with a network of non-interacting Ag cations, an unusual feature in argyrodite-like compounds. The characterization of Ag coordination geometries allows to confirm that the low-temperature structure and the room-temperature one are geometrically very similar to each other.

Supplementary Materials: The following are available online at https://www.mdpi.com/2075-163X/11/3/286/s1, file: Supplementary Materials CIF.

Author Contributions: The project was conceived by L.B., who also carried out the data collection, structure solution and refinement. L.B. wrote the paper with input from M.M. Both authors have read and agreed to the published version of the manuscript.

Funding: The research was funded by MIUR-PRIN2017, project "TEOREM deciphering geological processes using Terrestrial and Extraterrestrial ORE Minerals", prot. 2017AK8C32 (PI: Luca Bindi).

Data Availability Statement: Not applicable.

Acknowledgments: The authors give special thanks to the Editor of the Special Issue, Giovanni Ferraris, for the fruitful discussion on the twinning observed in spryite.

Conflicts of Interest: The authors declare no conflict of interest.

References

1. Bindi, L.; Keutsch, F.N.; Morana, M.; Zaccarini, F. Spryite, $Ag_8(As^{3+}, As^{5+}, Ge)S_6$: Structure determination and inferred absence of superionic conduction of the first As^{3+}-bearing argyrodite. *Phys. Chem. Miner.* **2017**, *44*, 75–82. [CrossRef]
2. Kuhs, W.F.; Nitsche, R.; Scheunemann, K. The argyrodites—A new family of tetrahedrally close-packed structures. *Mater. Res. Bull.* **1979**, *14*, 241–248. [CrossRef]
3. Evain, M.; Gaudin, E.; Boucher, F.; Petricek, V.; Taulelle, F. Structures and phase transitions of the A_7PSe_6 (A = Ag, Cu) argyrodite-type ionic conductors. I. Ag_7PSe_6. *Acta Cryst. B* **1998**, *54*, 376–383. [CrossRef]
4. Gu, X.; Watanabe, M.; Xie, X.; Peng, S.; Nakamuta, Y.; Ohkawa, M.; Hoshino, K.; Ohsumi, K.; Shibata, Y. Chenguodaite ($Ag_9FeTe_2S_4$): A new tellurosulfide mineral from the gold district of East Shandong Peninsula, China. *Chin. Sci. Bull.* **2008**, *53*, 3567–3573.
5. Gu, X.-P.; Watanabe, M.; Hoshino, K.; Shibata, Y. New find of silver tellurosulphides from the Funan gold deposit, East Shandong, China. *Eur. J. Miner.* **2003**, *15*, 147–155. [CrossRef]
6. Frank, D.; Gerke, B.; Eul, M.; Pöttgen, R.; Pfitzner, A. Synthesis and crystal structure determination of $Ag_9FeS_{4.1}Te_{1.9}$, the first example of an iron containing argyrodite. *Chem. Mater.* **2013**, *25*, 2339–2345. [CrossRef]
7. Bindi, L.; Biagioni, C. A crystallographic excursion in the extraordinary world of minerals: The case of Cu- and Ag-rich sulfosalts. *Acta Cryst. B* **2018**, *74*, 527–538. [CrossRef]
8. Eulenberger, G. Die Kristallstruktur Der Tieftemperaturmodifikation von Ag_8GeS_6. *Monatsh. Chem* **1977**, *108*, 901–913. [CrossRef]
9. Belin, R.; Aldon, L.; Zerouale, A.; Belin, C.; Ribes, M. Crystal structure of the non-stoichiometric argyrodite compound $Ag_{7-x}GeSe_5I_{1-x}$ (X=0.31). A highly disordered silver superionic conducting material. *Solid State Sci.* **2001**, *3*, 251–265. [CrossRef]
10. Oxford Diffraction. *CrysAlis RED, Version 1.171.31.2*; Oxford Diffraction Ltd.: Abington, UK, 2006.
11. Oxford Diffraction. *ABSPACK in CrysAlis RED*; Oxford Diffraction Ltd.: Abingdon, UK, 2006.
12. Sheldrick, G.M. A short history of SHELX. *Acta Cryst. A* **2008**, *64*, 112–122. [CrossRef]
13. Wilson, A.J.C. *International Tables for Crystallography: Mathematical, Physical, and Chemical Tables*; International Union of Crystallography: Chester, UK, 1992; Volume 3.
14. Nespolo, M.; Ferraris, G. Geminography—The science of twinning applied to the early-stage derivation of non-merohedric twin laws. *Krist. Cryst. Mater.* **2003**, *218*, 178–181. [CrossRef]
15. Petříček, V.; Dušek, M.; Palatinus, L. Crystallographic computing system JANA2006: General features. *Krist. Cryst. Mater.* **2014**, *229*, 345–352. [CrossRef]

6. Gaudin, E.; Petricek, V.; Boucher, F.; Taulelle, F.; Evain, M. Structures and phase transitions of the A_7PSe_6 (A = Ag, Cu) argyrodite-type ionic conductors. III. α-Cu_7PSe_6. *Acta Cryst. B* **2000**, *56*, 972–979. [CrossRef] [PubMed]
7. Bindi, L.; Downs, R.T.; Menchetti, S. The crystal structure of billingsleyite, $Ag_7(As,Sb)S_6$, a sulfosalt containg As_{5+}. *Canad. Miner.* **2010**, *48*, 155–162. [CrossRef]
8. Robinson, K.; Gibbs, G.V.; Ribbe, P.H. Quadratic elongation: A quantitative measure of distortion in coordination polyhedra. *Science* **1971**, *172*, 567–570. [CrossRef] [PubMed]

 *minerals*

Article

Polytypism of Compounds with the General Formula Cs{Al$_2$[TP_6O_{20}]} (T = B, Al): OD (Order-Disorder) Description, Topological Features, and DFT-Calculations

Sergey M. Aksenov [1,2,*], Alexey N. Kuznetsov [3,4], Andrey A. Antonov [1], Natalia A. Yamnova [5], Sergey V. Krivovichev [6,7] and Stefano Merlino [8]

[1] Laboratory of Nature-Inspired Technologies and Environmental Safety of the Arctic, Kola Science Centre, Russian Academy of Sciences, 14 Fersman str., 184209 Apatity, Russia; a.antonov@ksc.ru
[2] Geological Institute of Kola Science Centre, Russian Academy of Sciences, 14 Fersman Street, 184209 Apatity, Russia
[3] Faculty of Chemistry, Moscow State University, Vorobievy Gory, 119991 Moscow, Russia; alexei@inorg.chem.msu.ru
[4] Kurnakov Institute of General and Inorganic Chemistry RAS, Leninskii pr. 31, 119991 Moscow, Russia
[5] Faculty of Geology, Moscow State University, Vorobievy Gory, 119991 Moscow, Russia; natalia-yamnova@yandex.ru
[6] Nanomaterials Research Centre of Kola Science Centre, Russian Academy of Sciences, 14 Fersman Street, 184209 Apatity, Russia; skrivovi@mail.ru
[7] Department of Crystallography, Institute of Earth Sciences, Saint–Petersburg State University, University Emb. 7/9, 199034 St. Petersburg, Russia
[8] Accademia Nazionale dei Lincei, 00165 Rome, Italy; stefano.merlino38@gmail.com
* Correspondence: aks.crys@gmail.com

Citation: Aksenov, S.M.; Kuznetsov, A.N.; Antonov, A.A.; Yamnova, N.A.; Krivovichev, S.V.; Merlino, S. Polytypism of Compounds with the General Formula Cs{Al$_2$[TP_6O_{20}]} (T = B, Al): OD (Order-Disorder) Description, Topological Features, and DFT-Calculations. *Minerals* **2021**, *11*, 708. https://doi.org/10.3390/min11070708

Academic Editor: Giovanni Ferraris

Received: 12 June 2021
Accepted: 28 June 2021
Published: 30 June 2021

Publisher's Note: MDPI stays neutral with regard to jurisdictional claims in published maps and institutional affiliations.

Abstract: The crystal structures of compounds with the general formula Cs{$^{[6]}$Al$_2$[$^{[4]}TP_6O_{20}$]} (where T = Al, B) display order-disorder (OD) character and can be described using the same OD groupoid family. Their structures are built up by two kinds of nonpolar layers, with the layer symmetries $Pc(n)2$ (L_{2n+1}-type) and $Pc(a)m$ (L_{2n}-type) (category IV). Layers of both types (L_{2n} and L_{2n+1}) alternate along the **b** direction and have common translation vectors **a** and **c** ($a \sim 10.0$ Å, $c \sim 12.0$ Å). All ordered polytypes as well as disordered structures can be obtained using the following partial symmetry operators that may be active in the L_{2n} type layer: the 2_1 screw axis parallel to **c** [$- - 2_1$] or inversion centers and the 2_1 screw axis parallel to **a** [$2_1 - -$]. Different sequences of operators active in the L_{2n} type layer ([$- - 2_1$] screw axes or inversion centers and [$2_1 - -$] screw axes) define the formation of multilayered structures with the increased b parameter, which are considered as non-MDO polytypes. The microporous heteropolyhedral MT-frameworks are suitable for the migration of small cations such as Li$^+$, Na$^+$ Ag$^+$. Compounds with the general formula Rb{$^{[6]}M^{3+}$[$^{[4]}T^{3+}P_6O_{20}$]} (M = Al, Ga; T = Al, Ga) are based on heteropolyhedral MT-frameworks with the same stoichiometry as in Cs{$^{[6]}$Al$_2$[$^{[4]}TP_6O_{20}$]} (where T = Al, B). It was found that all the frameworks have common natural tilings, which indicate the close relationships of the two families of compounds. The conclusions are supported by the DFT calculation data.

Keywords: OD structures; polytypism; polymorphism; heteropolyhedral framework; modularity; topology; borophosphates; aluminophosphates; DFT

1. Introduction

Borophosphates (as well as borophosphate ceramics and glasses) attract interest because of their wide technological applications as materials with optical [1–5], electrochemical [6–9], magnetic [10–12], and catalytic [13–15] properties. Moreover, crystalline borophosphates and metal borophosphates with microporous structures are considered as zeolite-like materials [16–20]. Borophosphates are characterized by a wide diversity of tetrahedral and mixed triangular-tetrahedral anionic motifs [21–24], owing to the different possible coordination environments of boron. At present, more than 300 representatives of this class are known, which are characterized by anionic motifs with different dimensionalities (from isolated groups to 3D frameworks).

Compounds with the general formula $Cs\{^{[6]}Al_2[^{[4]}TP_6O_{20}]\}$ (where T = B [25], Al [26]) are based on microporous heteropolyhedral frameworks formed by tetrahedral borophosphate or aluminophosphate $[TP_6O_{20}]$-layers linked by isolated AlO_6 octahedra. The large framework cavities are filled by Cs^+ cations. As was previously shown, both $Cs\{Al_2[BP_6O_{20}]\}$ and $Cs\{Al_2[AlP_6O_{20}]\}$ are of modular character [27] and can be considered as polytypes belonging to the same OD family [20,25]. However, the corresponding groupoid family has not been reported so far.

In this paper we provide a complete OD-theoretical analysis of the compounds with the general formula $Cs\{^{[6]}Al_2[TP_6O_{20}]\}$ (where T = B [25], Al [26]) and derive symmetry and atom coordinates for the hypothetical MDO2 polytype. The energies of the observed and hypothetical structures of the family are calculated using the density functional theory (DFT). Possible ion-migration paths inside the microporous frameworks of the family are estimated for different alkaline ions using the topological analysis.

2. Methods

The symmetrical relations between the compounds have been analyzed using the OD theoretical approach [27–30] for the OD families containing more than one ($M > 1$) kind of layers [31]. The OD layers have been chosen in accordance with the equivalent region (ER) requirements [32]. As a reference structure for the further analysis, the MDO1 polytype observed in $Cs\{Al_2[AlP_6O_{20}]\}$ [26] was used. This compound was reported in the non-standard setting of the space group $C2cb$ [a = 10.0048(7) Å, b = 13.3008(10) Å, c = 12.1698(7) Å], which was transformed into the standard setting $Aea2$ using the [00–1/010/100] matrix (the resulting unit cell parameters are: a = 12.1698(7) Å b = 13.3008(10) Å, c = 10.0048(7) Å). The unit-cell parameters and space groups of the crystal structures of $Cs\{Al_2[BP_6O_{20}]\}$ polytypes have been transformed accordingly in order to preserve the orientation and stacking direction of the OD-layers.

Topological analysis of the frameworks was performed by means of natural tilings (the smallest polyhedral cationic clusters that form a framework) of the 3D cation nets [33]. The complexity parameters of the frameworks in different polytypes were calculated as Shannon information amounts per atom (I_G) and per reduced unit cell ($I_{G,total}$) [34,35]. To analyze the migration paths of alkaline cations in the structures, the Voronoi method [36], which has proven itself in the study of cationic conductors of various types [37,38], was used. Topological and complexity parameters for the whole structures as well as ion migration paths have been calculated using the ToposPro software [39].

DFT calculations on the existing MDO-, non-MDO-4O, as well as hypothetical MDO2 type polytypes (T = Al, B) were performed using the PBE exchange-correlation functional [40] of the GGA-type utilizing the projector augmented wave method (PAW) as implemented in the Vienna ab initio simulation package (VASP) [41,42]. The energy cut-off was set at 500 eV with a $10 \times 8 \times 8$ (MDO1, MDO2), and $6 \times 4 \times 4$ (non-MDO-4O) Monkhorst–Pack [43] k-point mesh used for Brillouin zone sampling. The convergence towards the k-point mesh was checked. Full optimization of the unit cell parameters and atomic coordinates was performed for all the structures except the MDO1 polytype of $Cs\{Al_2[BP_6O_{20}]\}$, for which the original cell parameters were retained and atomic coordinates optimized (as the compound was found to have the lowest energy, cell parameter optimization was deemed unnecessary). For the optimization, the structures were converted to the space group $P1$.

3. Results

3.1. OD (Order-Disorder) Relationships

The crystal structures of $Cs\{^{[6]}Al_2[^{[4]}TP_6O_{20}]\}$ (where T = B [25], Al [26]) belong to the same OD family of category IV [31] with two types of nonpolar OD layers and can be described by an OD groupoid [27]. The layers are as following:

1. Nonpolar L_{2n+1} type with the layer symmetry *pcn*2 [or *Pc*(*n*)2 in terms of the OD notation, where braces indicate the direction of missing periodicity [44]] was reported previously [20] and is represented by the tetrahedral $[^{[4]}TP_6O_{20}]$-layer (Figure 1);
2. Nonpolar L_{2n} type consists of aluminum and oxygen atoms on the borders of a thin slab with the layer symmetry *pcam* [*Pc*(*a*)*m* or $P2_1/c$ (2/*a*) $2_1/m$].

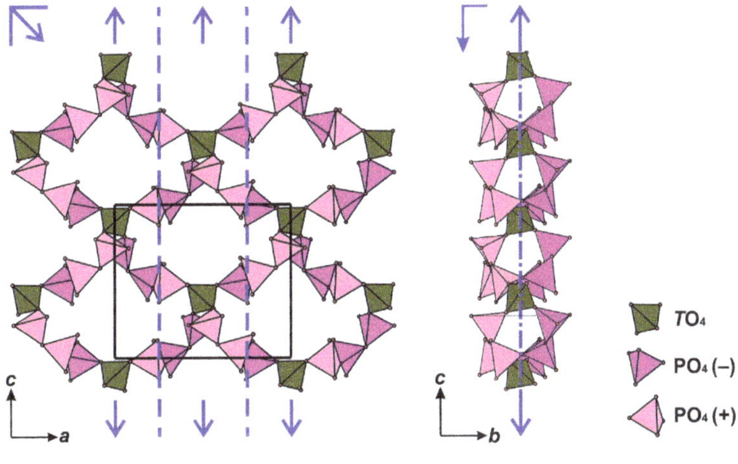

Symmetry of the *L₂ₙ₊₁* type layer – *Pc*(*n*)2

Figure 1. The general view of the tetrahedral L_{2n+1} type layer in the crystal structures of $Cs\{^{[6]}Al_2[^{[4]}TP_6O_{20}]\}$ polytypes. The fundamental building block (FBB) of the layer is represented by open-branched heptamer with the following descriptor [21,45]: 7□:[3□]2□|2□|□|□. Modified after [20].

Layers of both types (L_{2n} and L_{2n+1}) alternate along the **b** direction and have common translation vectors **a** and **c** ($a \sim 10.0$ Å, $c \sim 12.0$ Å), with b_0, the distance between the two nearest equivalent layers, corresponding to one half of the b parameter of the compound studied by Lesage et al. [26]. Because the symmetry of the L_{2n} type layers is higher than that of the L_{2n+1} type layers, polytypic relations are possible. All ordered polytypes as well as disordered structures can be obtained using the following symmetry operators that may be active in the L_{2n} type layer: the 2_1 screw axis parallel to **c** [– – 2_1] or inversion centers and the 2_1 screw axis parallel to **a** [2_1 – –] (Figure 2) [20]. The symmetry relation common to all polytypes of this family are described by the OD groupoid family symbol:

$$Pc(n)2 \qquad P\,2_1/c\,(2/a)\,2_1/m$$
$$[r,\,0] \tag{1}$$

where $r = 0$; the first line contains the layer-group symbols of the two constituting layers, while the second line indicates positional relations between the adjacent layers [46].

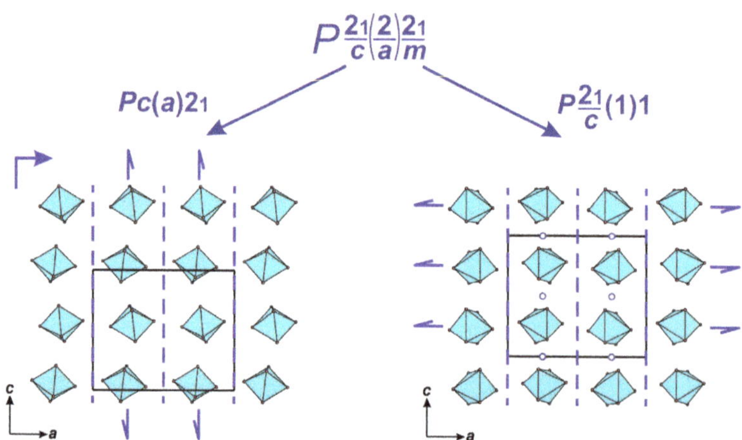

Figure 2. Different symmetry operators that may be active in the L_{2n} type layer: the 2_1 screw axis parallel to **c** $[-\,-\,2_1]$ (**left**) and inversion centers and the 2_1 screw axis parallel to **a** $[2_1\,-\,-]$ (**right**).

In accordance with the *NFZ* relation [27,28], there is only one kind of the $(L_{2n}, L_{2n+1}, L_{2n+2})$ triples and two kinds of the $(L_{2n-1}, L_{2n}, L_{2n+1})$ triples. Consequently, the smallest possible number of different triples in a structure is two and only two MDO polytypes are possible:

$$
\begin{array}{lcccc}
\text{OD} - \text{layer} & \text{Layer group} & \text{Subgroup of }\lambda\text{-}\tau\text{-operations} & N & F & Z \\
A^1 = L_{2n} & P\,2_1/c\,2/a\,2_1/m & P\,c\,2\,m & 4\searrow & \nearrow 1 \\
& \text{Symmetry of a layer pair} \rightarrow & P\,c\,1\,1 & 2 \\
A^2 = L_{2n+1} & P\,c\,n\,2 & P\,c\,1\,1 & 2\nearrow & \searrow 2
\end{array}
\tag{2}
$$

The first MDO structure (MDO1 polytype) (Figure 3, left) can be obtained when the $[-\,-\,2_1]$ operator is active in L_{2n} type layer. Through the action of this operator the asymmetric unit at x, y, z (I) is converted into the asymmetric unit at $-x, \frac{1}{2}-y, \frac{1}{2}+z$ (II); the latter unit is converted by the $[-\,-\,2]$ operator in the L_{2n+1} layer into the asymmetric unit at $x, \frac{1}{2}+y, \frac{1}{2}+z$ (III). I and III are related by the translation vector $\mathbf{t} = \mathbf{b}_0 + \mathbf{c}/2$, which is the generating operation, giving rise by the continuation to an A-centered structure with the basis vectors $\mathbf{a}, \mathbf{b} = 2\mathbf{b}_0, \mathbf{c}$ and the space group $Aea2$. The MDO1 polytype corresponds to the structure of $Cs\{Al_2[AlP_6O_{20}]\}$ with the following unit cell parameters: $a = 12.1698(7)$ Å, $b = 13.3008(10)$ Å, $c = 10.0048(7)$ Å [26].

The second MDO structure (MDO2 polytype) (Figure 3, right) can be obtained when the inversion centers and $[2_1\,-\,-]$ operators are both active in the L_{2n} type layer. Through the action of the operator $[2_1\,-\,-]$ the asymmetric unit at x, y, z (I) is converted into the asymmetric unit $\frac{1}{2}+x, -y, \frac{1}{2}-z$ (II); the latter unit is converted by the $[-\,n\,-]$ operator in the L_{2n+1} layer into the asymmetric unit $x, \frac{1}{2}+y, -z$ (III); (I) and (III) are related by a b glide normal to **c**, with translational component b_0, which is the generating operation: its continuation also generates an orthorhombic structure with the basis vectors $\mathbf{a}, \mathbf{b} = 2\mathbf{b}_0, \mathbf{c}$ (the same for the MDO1 polytype) and the space group $Pcnb$ (or $Pbcn$ in the standard setting). The MDO2 polytype has not yet been observed for the compound with the general formula $Cs\{^{[6]}Al_2[^{[4]}TP_6O_{20}]\}$. The calculated atomic coordinates for the MDO2 polytype are given in Table S1 (Supplement Materials).

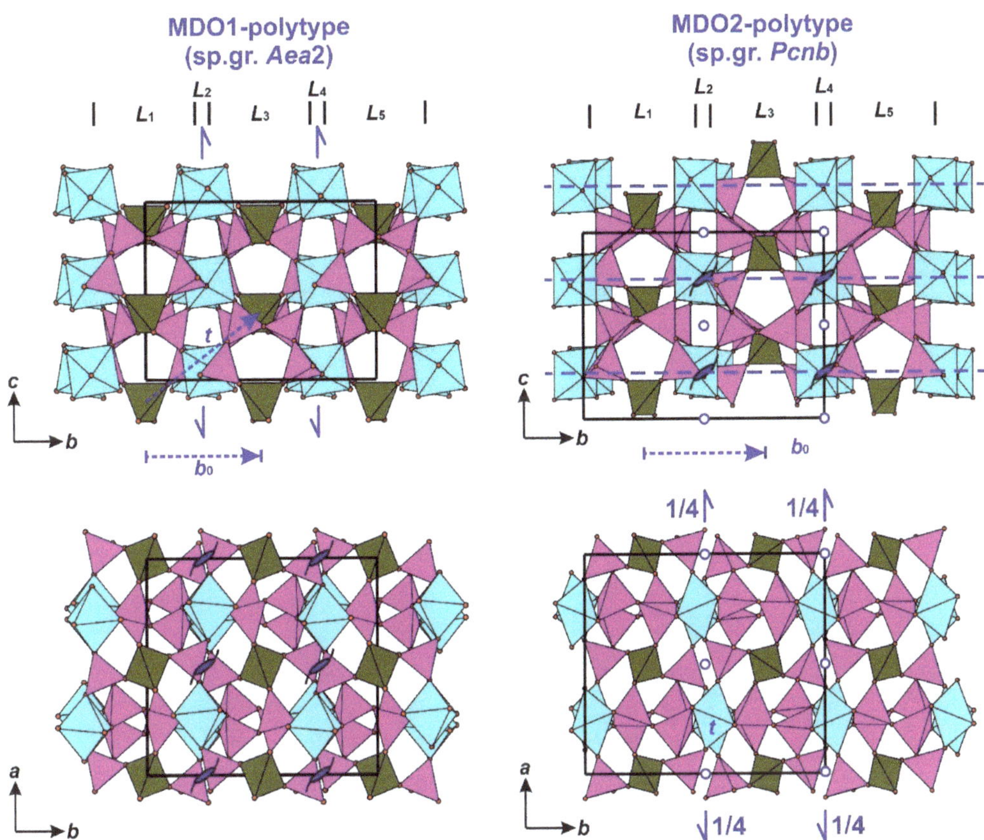

Figure 3. The general views of the MDO1 (with the space group *Aea*2) and MDO2 (with the space group *Pcnb*) polytypes. The operations active in the L_{2n} type layers as well as the generating operations are shown. Legend: AlO_6-octahedra are colored in cyan; PO_4-tetrahedra are colored in purple; TO_4-tetrahedra are colored in dark yellow. Extra-frameworks Cs atoms are omitted.

Different sequences of operators active in the L_{2n} type layer ([– – 2_1] screw axes or inversion centers and [2_1 – –] screw axes) define the formation of structures with the increased b parameter, which are considered as non-MDO polytypes (because of the presence of more than one kind of (L_{2n-1}, L_{2n}, L_{2n+1}) triples) [27]. The compound Cs{Al₂[BP₆O₂₀]} [25] contains four L_{2n} and L_{2n+1} types layers, where each L_{4n} type layer has active [2_1 – –] screw axes, while in the L_{4n+2} type the inversion centers and [– – 2_1] screw axes are active (Figure 4). The AlO_6 octahedra in the L_{2n+2} and L_{2n+4} type layers are tilted slightly differently, which can be explained by the "desymmetrization" effect of OD structures [27,47,48], when the ideal symmetry suffers slight (in some cases severe) distortions and the symmetry of OD layers in the polytype is lower than the idealized one. The orthorhombic structure of Cs{Al₂[BP₆O₂₀]}–4O is characterized by the basis vectors **a**, **b** = 4**b₀**, **c** (where $a = 11.815(2)$ Å, $b = 26.630(4)$ Å, $c = 10.042(2)$ Å [25]) and the space group *Pcab* (nonstandard setting of the space group *Pbca*).

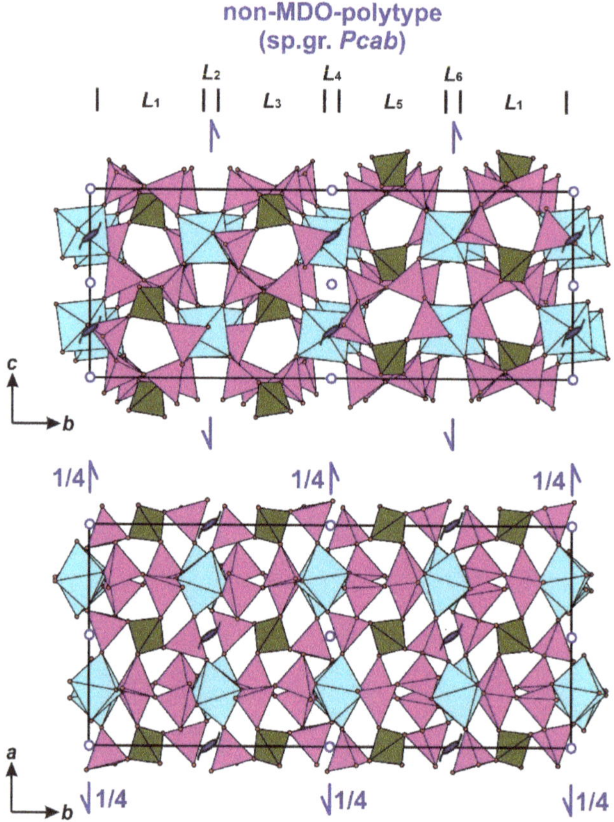

Figure 4. The general views of the non-MDO 4*O* polytype. The operations active in the L_{2n} type layers are shown. Modified after [20].

3.2. Topological Features

Compounds with the general formula $Cs\{^{[6]}Al_2[^{[4]}TP_6O_{20}]\}$ (where T = B [25], Al [26]) are characterized by the heteropolyhedral MT-frameworks [20,49–51] of MO_6-octahedra and TO_4-tetrahedra related to classic zeolites and zeolite-type materials where all oxygen ligands are bridged between two cations only [52]. In accordance with the theory of mixed anionic radicals [53–55], the general crystal chemical formula of the framework (taking into account the degree of sharing of oxygen ligands) can be written as [20]:

$$\left\{ M_m \left[(T_1)_{n_1}(T_2)_{n_2} O_{3m+2(n_1+n_2)} \right] \right\}^{m(V_M-6)+n_1(V_{T_1}-4)+n_2(V_{T_2}-4)}, \tag{3}$$

where where m and n_i, V_M and V_{T_i} are the valences of the M and T_i cations, respectively. If $M = M^{3+}$, $T_1 = T^{3+}$, $T_2 = P^{5+}$, $m = z$, $n_1 = y$, $n_2 = z$, the Formula (3) can be rewritten as:

$$\left\{ M_x \left[T_y P_z O_{3x+2(y+z)} \right] \right\}^{-3x-y+z}. \tag{4}$$

Taking into account the observed ratio between the x, y, and z coefficients, the stoichiometry of the heteropolyhedral MT-framework is:

$$\{M_2[TP_6O_{20}]\}^{1-}. \tag{5}$$

Topological features of the MDO1 and non-MDO 4*O* polytypes have been described previously [20]. The cationic 3D net corresponding to the heteropolyhedral *MT*-framework of MDO2 polytype consists of four natural tiles (Figure 5): $[4.6^2]_2[3.5.6^2]_2[4^4.5^2.7^2][3^2.4^2.6^6.7^2]$. The (6*T*1*M*)-$[4.6^2]$ and (6*T*2*M*)-$[3.5.6^2]$ tiles are topologically equal to those observed in the MDO1 and non-MDO 4*O* polytypes; the (10*T*4*M*)-$[4^4.5^2.7^2]$ tile is equal to that in the non-MDO 4*O* polytype. The (16*T*6*M*)-$[3^2.4^2.6^6.7^2]$ tile is unique and is present in the MDO2 polytype only (Table 1).

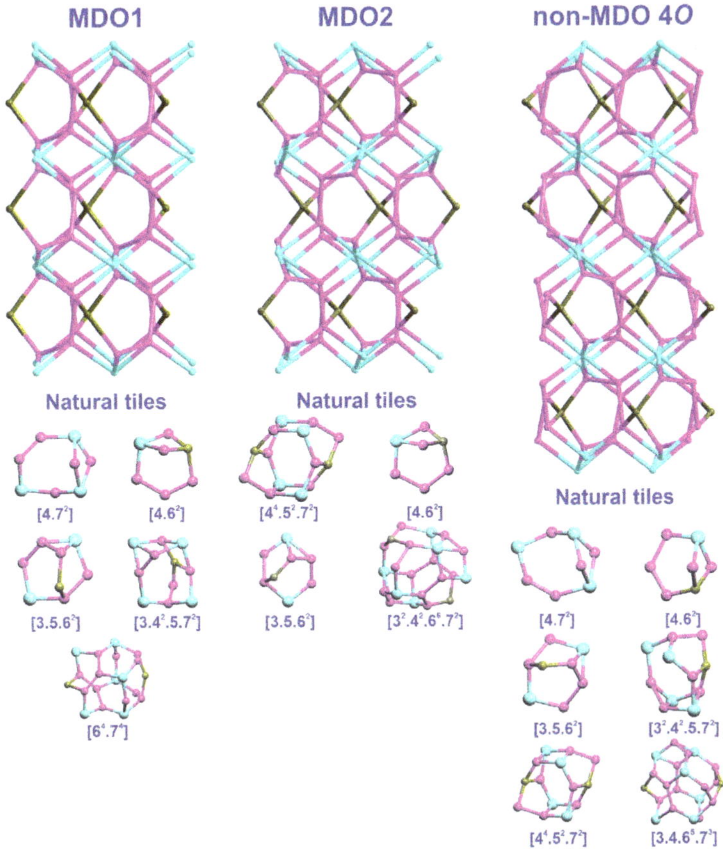

Figure 5. Topological features of the 3D cationic nets of the heteropolyhedral *MT*-frameworks in the structures of MDO1, MDO2 and non-MDO 4*O* polytypes of compounds with the general formula $Cs\{^{[6]}Al_2[^{[4]}TP_6O_{20}]\}$ (where T = B, Al).

Table 1. The natural tiles in the *MT*-frameworks of the polytypes of compounds with the general formula $Cs\{^{[6]}Al_2[^{[4]}TP_6O_{20}]\}$ (where T = B, Al).

Polytype	Natural Tiles					
MDO1	$[4.6^2]_2$	$[3.5.6^2]_2$	$[6^4.7^4]$	$[3.4^2.5.7^2]_2$	$[4.7^2]_2$	
MDO2	$[4.6^2]_2$	$[3.5.6^2]_2$	$[4^4.5^2.7^2]$	$[3^2.4^2.6^6.7^2]$		
non-MDO 4*O*	$[4.6^2]_4$	$[3.5.6^2]_4$	$[4^4.5^2.7^2]$	$[3.4^2.5.7^2]_2$	$[4.7^2]_2$	$[3.4.6^5.7^3]_2$

Note. The point symbol of the 3D net has the form $A^a.\ B^b\ \ldots$ indicating that there are a angles with shortest cycles that are A-cycles, b angles with shortest cycles that are B-cycles, etc., with $A < B,< \cdots$ and $a + b + \cdots = n(n-1)/2$ [33]. The topologically equivalent tiles are colored in the same color.

The complexity parameters of the heteropolyhedral *MT*-framework of MDO2 polytype are: v = 116 atoms; I_G = 3.892 bits/atom; $I_{G,total}$ = 451.526 bits/unit cell. The complexity parameters increase in the row MDO1 → MDO2 → non-MDO 4*O*.

3.3. Ion Migration Path

Migration maps of Na$^+$ cation were constructed for the MDO1, MDO2, and non-MDO 4*O* polytypes (Table 2). Despite the presence of large pores filled by large Cs$^+$ ions, the size of the effective windows between them is not enough for the migration of large alkaline cations. However, all the types of the microporous heteropolyhedral *MT*-framework are suitable for the migration of smaller ions such as Li$^+$, Na$^+$ Ag$^+$. The types of migration maps depend on the topological type of the MT-framework (Figure 6), in particular, for Na ions, the maps are represented by 2D layers parallel to (100) for the MDO1 and non-MDO 4*O* polytypes, while for the MDO2 polytype it is represented by the system of parallel 1D channels directed along [010] (Figure 6). In the case of Li$^+$ ions, the migration 3D maps are similar for all the types of the frameworks.

Table 2. The natural tiles in the *MT*-frameworks of the polytypes of compounds with the general formula Cs{$^{[6]}$Al$_2$[$^{[4]}$TP$_6$O$_{20}$]} (where T = B, Al).

Polytype	Natural Tiles					
	Li$^+$	Na$^+$	Ag$^+$	K$^+$	Rb$^+$	Cs$^+$
MDO1	3D	2D	2D	–	–	–
MDO2	3D	1D	1D	–	–	–
non-MDO 4*O*	3D	2D	2D	–	–	–

Note: The following significance criteria for elementary channels (R_{chan}) and voids (R_{sd}) have been used for the construction of migration maps: Li$^+$ (R_{chan} = 2.02 Å; R_{sd} = 1.38 Å); Na$^+$ (R_{chan} = 2.16 Å; R_{sd} = 1.54 Å); Ag$^+$ (R_{chan} = 2.20 Å; R_{sd} = 1.58 Å), K$^+$ (R_{chan} = 2.30 Å; R_{sd} = 1.70 Å); Rb$^+$ (R_{chan} = 2.38 Å; R_{sd} = 1.78 Å); Cs$^+$ (R_{chan} = 2.47 Å; R_{sd} = 1.88 Å).

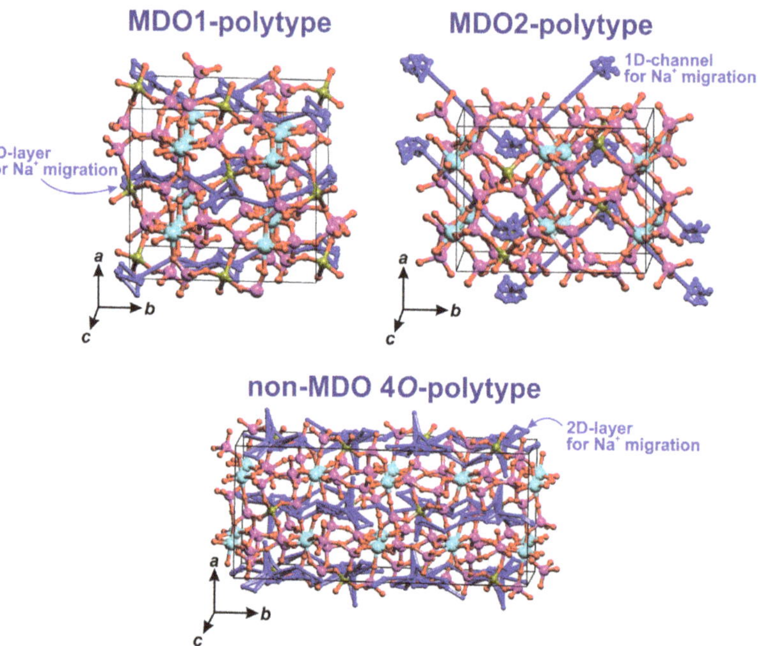

Figure 6. Possible ion migration path of Na$^+$ cations in the crystal structures of Cs{$^{[6]}$Al$_2$[$^{[4]}$TP$_6$O$_{20}$]} polytypes.

3.4. DFT Calculations

In order to gain more insight into the stability of various polytypes, energy-wise, we have performed DFT calculations on the existing as well as hypothetical compounds with the general formula $Cs\{Al_2[TP_6O_{20}]\}$ (T = Al, B) with the structures belonging to MDO1, MDO2, and non-MDO 4O type polytypes, for T = Al; B. The comparative data and optimized unit cell parameters are given in Table 3 (for MDO1, T = Al, original unit cell metrics were retained).

Table 3. Comparative data for the frameworks of different polytypes.

Parameter	MDO1 Polytype		MDO2 Polytype		Non-MDO 4O Polytype	
	T = B	T = Al	T = B	T = Al	T = B	T = Al
Unit cell parameters (Å), a, b, c	n.d.	12.170, 13.301, 10.005	n.d.	n.d.	11.815, 26.630, 10.042	n.d.
Volume (Å^3)	n.d.	1619.46	n.d.	n.d.	3159.55	n.d.
Optimized unit cell parameters (Å), a, b, c	12.0296, 13.2109, 9.9017	12.1698,* 13.3008, * 10.0048 *	11.7893, 13.4876, 10.1609	11.9479, 13.6593, 10.3157	11.8248, 26.7192, 10.0423	12.2217, 26.9351, 10.1760
Optimized volume (Å^3)	1573.60	1619.46	1615.68	1683.52	3172.86	3349.86
Z	4		4		8	
Energy per formula unit (eV)	−219.1885	−218.2701	−219.2479	−217.7659	−219.2780	−218.2109
FD [$(M + T)/1000$ Å^3]		19.76	19.81	19.01	22.69	21.49
v (atoms), framework, all	58, 60		116, 120		232, 240	
I_G (bits/atom), framework, all	3.892, 3.974		3.892, 3.974		4.858, 4.907	
$I_{G,\,total}$ (bits/unit cell), framework, all	225.763, 238.413		451.526, 476.827		1127.052, 1177.654	

n.d.—no data, because of the absence of structural information; original unit cell parameter.

As seen from the comparison between the original and optimized cells of $Cs\{Al_2[BP_6O_{20}]\}$ of the non-MDO 4O type, they are in a very good agreement, with the difference in volume of ca. 13 Å, i.e., ca. 0.4% (see Table 3). The optimized coordinates in all structures showed only minimal shifts from their original positions, mostly associated with a very small rotation of tetrahedra. It is important to note that, despite unconstrained optimization, all the structures, observed as well as hypothetical, retained their original cell symmetries.

As seen from Table 3, for the T = Al series, the structure with the lowest energy was the MDO1-type polytype. However, the non-MDO 4O-type structure was only ca. 0.06 eV higher in energy, which corresponds to ca. 6.2 kJ/mol. This difference is not large, yet is arguably outside the margin of error for the computational method used, which is commonly estimated as 1–2 kJ/mol. The important thing here is that both experimentally observed types of structures (albeit not both of them for T = Al), showed comparable energies. Moreover, our calculations indicate that, under the right conditions, it might be possible to obtain the non-MDO 4O polytype for aluminum. Regarding the MDO2-type structure, the optimization gave us a stable minimum structure with the energy of ca. 0.5 eV (ca. 49 kJ/mol) higher than MDO1. This means that, potentially, such a structure might exist, however, the energy difference to the lowest energy structure is significant, and thus it might be difficult to stabilize such a polytype.

For the T = B series, once again the lowest energy corresponds to the experimentally observed structure, this time it is the non-MDO 4O polytype (see Table 3). In this case, however, its energy is only ca. 0.03 eV (ca. 3 kJ/mol) lower than that of the hypothetical MDO2-type structure. The difference is on the border of the perceived accuracy of the computational method, thus the MDO2 polytype appears to be a good candidate for the experimental discovery. The MDO1-type structure in this case looks like the least favorable, energy-wise, with the difference between its energy and minimal structure being ca. 0.09 eV (ca. 8.6 kJ/mol). This is clearly outside the margin of error; however, the difference is small enough to be compensated by various effects in real crystals. It must also be noted regarding all our calculations, that by their very nature they simulate ideal periodic crystals in their ground state at 0 K. In addition, in our computations we cannot account for potential kinetic hindrance of certain paths of compound formation.

4. Discussion

The heteropolyhedral *MT*-frameworks with similar stoichiometry (3) have been found in compounds with the general formula $Rb\{^{[6]}M^{3+}{}_2[^{[4]}T^{3+}P_6O_{20}]\}$, where M = Al, Ga; T = Al, Ga [26,56]. The unit cell parameters are similar to those for MDO1 and MDO2 polytypes of $Cs\{Al_2[TP_6O_{20}]\}$ (T = Al, B): a = 9.876–10.002 Å; b = 12.885–13.082 Å; c = 11.919–12.071 Å; space group $C222_1$. Their crystal structures contain mixed tetrahedral $[TP_6O_{20}]$-chains, which are linked by the MO_6-octahedra (Figure 7). The tetrahedral chain is formed by the condensation of FBU, an open-branched heptamer with the descriptor 7□:[3□]2□|2□|□|□ similar to that for the tetrahedral $[TP_6O_{20}]$-layers in $Cs\{Al_2[TP_6O_{20}]\}$ (T = Al, B). The negative charge of the framework is balanced by Rb^+ ions.

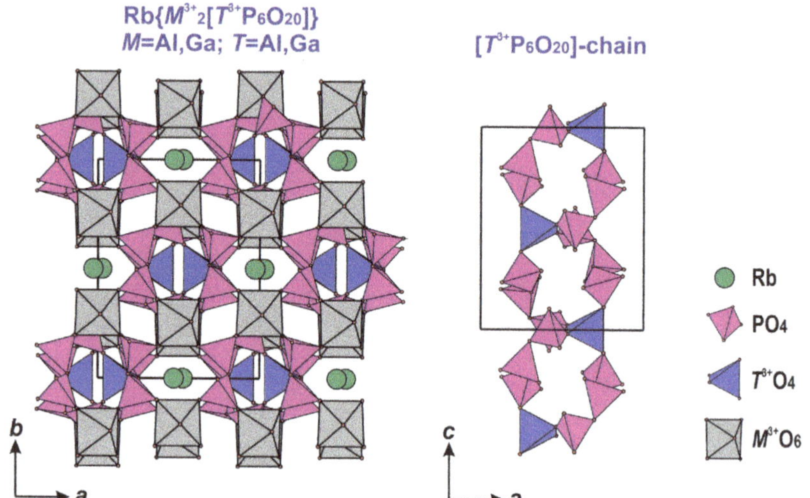

Figure 7. The general view of the crystal structure of compounds with the general formula $Rb\{^{[6]}M^{3+}{}_2[^{[4]}T^{3+}P_6O_{20}]\}$ (where M = Al, Ga; T = Al, Ga [26,51]) and a tetrahedral chain going along [001].

Despite of the absence of the tetrahedral layers, the *MT*-framework can also be considered as the result of alternation along **b** of two types of nonpolar OD layers parallel to (010):

1. The first one corresponds to a layer with the symmetry $P2(2)2_1$ consisting of tetrahedral chains. The tetrahedral layer in $Cs\{Al_2[TP_6O_{20}]\}$ and tetrahedral pseudolayer in $Rb\{M_2[TP_6O_{20}]$ are formed by the same FBU and demonstrate the symmetrical relationship (Figure 8) indicating the possible OD-character as was previously shown for compounds with tetrameric [57] and pentameric [20] borophosphate FBUs, as well as for the silicate layers [58,59];

2. The second one consists of an octahedral layer with the symmetry $P2_1(2)2_1$ similar to that observed in $Cs\{Al_2[TP_6O_{20}]\}$ (T = Al, B) (the layer group $P2_122_1$ is a subgroup of the layer group $Pcam$). To date, there are no other polytypes of this type of framework, however they may be found later.

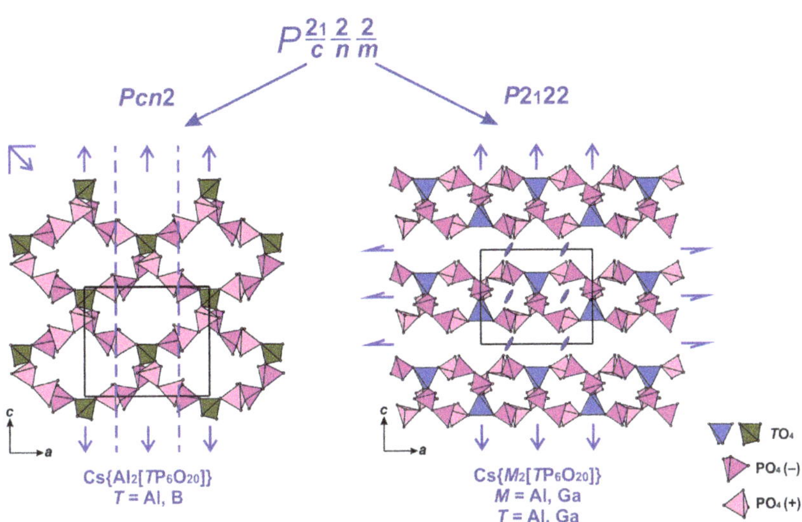

Figure 8. The symmetrical relationship between tetrahedral layers and tetrahedral pseudolayers in compounds with the general formulas Cs{Al₂[TP_6O_{20}]} (T = Al, B) and Rb{M_2[TP_6O_{20}]} (M = Al, Ga; T = Al, Ga), respectively. The orientation of the tetrahedral pseudolayer (right) have been changed using the [001/010/100] matrix in comparison with that in the crystal structures.

Topological features of the MT-framework are reflected in the sequence of its natural tiles: $[4.6^2]_2[4.7^2]_2[3.5.6^2]_2[3.4^2.5.7^2]_2[6^4.7^4]$. It should be noted that three tilings ($[4.6^2]$, $[4.7^2]$), and $[3.5.6^2]$) are topologically equivalent to those in the Cs{Al₂[TP_6O_{20}]} (T = Al, B) compounds, which indicate the relation of the two types of the {$^{[6]}M^{3+}{}_2[^{[4]}T^{3+}P_6O_{20}]$}-frameworks.

5. Conclusions

The polytypism of compounds with the general formula Cs{Al₂[TP_6O_{20}]} (T = Al, B) has been described using the OD theory approach. The crystal structure of the hypothetical MDO2 polytype has been proposed and optimized using DFT calculations. It was shown that the heteropolyhedral MT-frameworks of all the polytypes contain similar natural tilings. The compounds with the general formula Rb{$^{[6]}M^{3+}{}_2[^{[4]}T^{3+}P_6O_{20}]$} ($M$ = Al, Ga; T = Al, Ga) have the heteropolyedral MT-frameworks with the same stoichiometry. It was found that all the frameworks had common natural tilings, which indicates the relationship of both families of compounds. Our computational data agree well with those which are experimentally available and, we believe, provide a reasonable basis for an internally consistent picture which supports crystallographic considerations concerning the formation of the polytypes of compounds with the general formula Cs{Al₂[TP_6O_{20}]} (T = Al, B). Thus, it is seems possible to synthesize the MDO2 polytype as well as the "missing" members, such as MDO1 polytype of Cs{Al₂[BP₆O₂₀]} and non-MDO 4O polytype of Cs{Al₂[AlP₆O₂₀]} using hydrothermal techniques.

Supplementary Materials: The following are available online at https://www.mdpi.com/article/10.3390/min11070708/s1, Table S1: Site coordinates (*xyz*) and site multiplicities (Mult.) for MDO2 polytype of Cs{Al₂[TP_6O_{20}]}. The optimized unit cell parameters and atomic coordinates for MDO1, MDO2, and non-MDO-4O polytypes of compounds with the general formula Cs{Al₂[TP_6O_{20}]} (T = Al, B) are given (the atomic coordinates are given for the whole crystal structures for the space group *P*1).

Author Contributions: Conceptualization, S.M.A., A.N.K. and S.M.; formal analysis, S.M.A., A.A.A. and N.A.Y.; writing—review and editing, S.M.A. and S.M.; supervision, A.N.K., S.V.K. and S.M. All authors have read and agreed to the published version of the manuscript.

Funding: This research was funded by the Russian Science Foundation (Project No. 20-77-10065) (S.M.A., A.A.A.).

Acknowledgments: The authors are grateful to reviewers and the Special Issue editor Professor Giovanni Ferraris for their useful comments and suggestions.

Conflicts of Interest: The authors declare no conflict of interest.

References

1. Vinodkumar, P.; Panda, S.; Jaiganesh, G.; Padhi, R.K.; Madhusoodanan, U.; Panigrahi, B.S. SrBPO5: Ce^{3+}, Dy^{3+}—A cold white-light emitting phosphor. *Spectrochim. Acta Part. A Mol. Biomol. Spectrosc.* **2021**, *253*, 119560. [CrossRef]

2. He, X.; Hu, D.; Yang, G.; Adamietz, F.; Rodriguez, V.; Dussauze, M.; Fargues, A.; Fargin, E.; Cardinal, T. Microstructured SHG patterns on Sm_2O_3-doped borophosphate niobium glasses by laser-induced thermal poling. *Ceram. Int.* **2021**, *47*, 10123–10129. [CrossRef]

3. Joseph, P.A.J.; Maheshvaran, K.; Rayappan, I.A. Structural and optical studies on Dy^{3+} ions doped alkali lead borophosphate glasses for white light applications. *J. Non. Cryst. Solids* **2021**, *557*, 120652. [CrossRef]

4. Xiang, J.; Fang, Z.; Yang, D.; Zheng, Y.; Zhu, J. Optimizational orange emitting behavior of Li_2Na_1-BP_2O_8:xPr solid solutions under an short-wave ultraviolet irradiation. *Scr. Mater.* **2020**, *187*, 82–87. [CrossRef]

5. Zhao, D.; Cheng, W.-D.; Zhang, H.; Huang, S.-P.; Xie, Z.; Zhang, W.-L.; Yang, S.-L. $KMBP_2O_8$ (M = Sr, Ba): A New Kind of Noncentrosymmetry Borophosphate with the Three-Dimensional Diamond-like Framework. *Inorg. Chem.* **2009**, *48*, 6623–6629. [CrossRef]

6. Magistris, A.; Chiodelli, G.; Duclot, M. Silver borophosphate glasses: Ion transport, thermal stability and electrochemical behaviour. *Solid State Ion.* **1983**, *9–10*, 611–615. [CrossRef]

7. Mouyane, M.; Jumas, J.-C.; Olivier-Fourcade, J.; Cassaignon, S.; Jordy, C.; Lippens, P.-E. One-pot synthesis of tin-borophosphate carbon composites as anode materials for Li-ion batteries. *J. Solid State Chem.* **2016**, *233*, 52–57. [CrossRef]

8. Yaghoobnejad Asl, H.; Stanley, P.; Ghosh, K.; Choudhury, A. Iron Borophosphate as a Potential Cathode for Lithium- and Sodium-Ion Batteries. *Chem. Mater.* **2015**, *27*, 7058–7069. [CrossRef]

9. Shenouda, A.Y.; Liu, H.K. Electrochemical behaviour of tin borophosphate negative electrodes for energy storage systems. *J. Power Sources* **2008**, *185*, 1386–1391. [CrossRef]

10. Shvanskaya, L.; Yakubovich, O.; Krikunova, P.; Ovchenkov, E.; Vasiliev, A. Chain caesium borophosphates with B:P ratio 1:2 Synthesis, structure relationships and low-temperature thermodynamic properties. *Acta Crystallogr. Sect. B Struct. Sci. Cryst. Eng. Mater.* **2019**, *75*, 1174–1185. [CrossRef] [PubMed]

11. Yakubovich, O.V.; Shvanskaya, L.V.; Kiriukhina, G.V.; Volkov, A.S.; Dimitrova, O.V.; Ovchenkov, E.A.; Tsirlin, A.A.; Shakin, A.A.; Volkova, O.S.; Vasiliev, A.N. Crystal structure and spin-trimer magnetism of $Rb_{2.3}(H_2O)_{0.8}Mn_3[B_4P_6O_{24}(O,OH)_2]$. *Dalt. Trans* **2017**, *46*, 2957–2965. [CrossRef]

12. Shvanskaya, L.; Yakubovich, O.; Melchakova, L.; Ivanova, A.; Vasiliev, A. Crystal chemistry and physical properties of the $A_2M_3(H_2O)_2[B_4P_6O_{24}(OH)_2]$ (A = Cs, Rb; M = Ni, Cu, (Ni, Fe)) borophosphate family. *Dalt. Trans.* **2019**, *48*, 8835–8842. [CrossRef] [PubMed]

13. Scheide, M.R.; Peterle, M.M.; Saba, S.; Neto, J.S.S.; Lenz, G.F.; Cezar, R.D.; Felix, J.F.; Botteselle, G.V.; Schneider, R.; Rafique, J.; et al Borophosphate glass as an active media for CuO nanoparticle growth: An efficient catalyst for selenylation of oxadiazoles and application in redox reactions. *Sci. Rep.* **2020**, *10*, 15233. [CrossRef] [PubMed]

14. Matzkeit, Y.H.; Tornquist, B.L.; Manarin, F.; Botteselle, G.V.; Rafique, J.; Saba, S.; Braga, A.L.; Felix, J.F.; Schneider, R. Borophosphate glasses: Synthesis, characterization and application as catalyst for bis(indolyl)methanes synthesis under greener conditions. *J. Non. Cryst. Solids* **2018**, *498*, 153–159. [CrossRef]

15. Wang, B.; Lu, W.-X.; Huang, Z.-Q.; Chen, W.-J.; Xie, J.-L.; Pan, D.-S.; Zhou, L.-L.; Song, J.-L. Amorphous N-Doped Cobalt Borophosphate Nanoparticles as Robust and Durable Electrocatalyst for Water Oxidation. *ACS Sustain. Chem. Eng.* **2019**, *7*, 13981–13988. [CrossRef]

16. Belokoneva, E.L.; Dimitrova, O.V. $Fe_{2.5}[BP_2O_7(OH)_2][PO_3(OH)][PO_3(O_{0.5}OH_{0.5})] \cdot H_2O$, a new phosphate-borophosphate with a microporous structure. *Crystallogr. Rep.* **2015**, *60*, 361–366. [CrossRef]

17. Yang, M.; Yan, P.; Xu, F.; Ma, J.; Welz-Biermann, U. Role of boron-containing ionic liquid in the synthesis of manganese borophosphate with extra-large 16-ring pore openings. *Microporous Mesoporous Mater.* **2012**, *147*, 73–78. [CrossRef]

18. Kang, Q.-Y.; Song, Q.; Li, S.-Y.; Liu, Z.-H. Thermodynamic properties of microporous materials for two borophosphates, $K[ZnBP_2O_8]$ and $NH4[ZnBP_2O_8]$. *J. Chem. Thermodyn.* **2014**, *69*, 43–47. [CrossRef]

19. Yang, T.; Li, G.; Ju, J.; Liao, F.; Xiong, M.; Lin, J. A series of borate-rich metalloborophosphates $Na_2[M^{II}B_3P_2O_{11}(OH)]$ $0.67H_2O$ (M^{II}=Mg, Mn, Fe, Co, Ni, Cu, Zn): Synthesis, structure and magnetic susceptibility. *J. Solid State Chem.* **2006**, *179*, 2534–2540. [CrossRef]

20. Aksenov, S.M.; Yamnova, N.A.; Borovikova, E.Y.; Stefanovich, S.Y.; Volkov, A.S.; Deyneko, D.V.; Dimitrova, O.V.; Hixon, A.E.; Krivovichev, S.V. Topological features of borophosphates with mixed frameworks. Synthesis, crystal structure of $Li_3\{Al_2[BP_4O_{16}]\}\cdot 2H_2O$, and comparative crystal chemistry. *J. Struct. Chem.* **2020**, *61*. [CrossRef]

21. Ewald, B.; Huang, Y.-X.; Kniep, R. Structural Chemistry of Borophosphates, Metalloborophosphates, and Related Compounds. *Z. Anorg. Allg. Chem.* **2007**, *633*, 1517–1540. [CrossRef]

22. Gurbanova, O.A.; Belokoneva, E.L. Comparative crystal chemical analysis of borophosphates and borosilicates. *Crystallogr. Rep.* **2007**, *52*, 624–633. [CrossRef]

23. Li, M.; Verena-Mudring, A. New Developments in the Synthesis, Structure, and Applications of Borophosphates and Metalloborophosphates. *Cryst. Growth Des.* **2016**, *16*, 2441–2458. [CrossRef]

24. Yakubovich, O.; Steele, I.; Massa, W. Genetic aspects of borophosphate crystal chemistry. *Z. Krist. Cryst. Mater.* **2013**, *228*. [CrossRef]

25. Shvanskaya, L.V.; Yakubovich, O.V.; Belik, V.I. New type of borophosphate anionic radical in the crystal structure of CsAl2BP6O20. *Crystallogr. Rep.* **2016**, *61*, 786–795. [CrossRef]

26. Lesage, J.; Guesdon, A.; Raveau, B. Two aluminotriphosphates with closely related intersecting tunnel structures involving tetrahedral "AlP" chains and layers: AAl$_3$(P$_3$O$_{10}$)$_2$, A=Rb, Cs. *J. Solid State Chem.* **2005**, *178*, 1212–1220. [CrossRef]

27. Ferraris, G.; Makovicky, E.; Merlino, S. *Crystallography of Modular Materials*; Oxford University Press: Oxford, UK, 2008; ISBN 9780191712111.

28. Dornberger-Schiff, K. Grundzüge einer Theorie der OD-Strukturen aus Schichten. *Abh. Dtsch. Akad. Wiss. Berlin. Kl. Chem. Geol. Biol.* **1964**, *3*, 1–107.

29. Dornberger-Schiff, K. *Lehrgang Über OD-Strukturen*; Akademie-Verlag: Berlin, Germany, 1966.

30. Dornberger-Schiff, K.; Grell-Niemann, H. On the theory of order–disorder (OD) structures. *Acta Crystallogr.* **1961**, *14*, 167–177. [CrossRef]

31. Dornberger-Schiff, K.; Grell, H. Geometrical properties of MDO polytypes and procedures for their derivation. II. OD families containing OD layers of M > 1 kinds and their MDO polytypes. *Acta Crystallogr. Sect. A* **1982**, *38*, 491–498. [CrossRef]

32. Grell, H. How to choose OD layers. *Acta Crystallogr. Sect. A Found. Crystallogr.* **1984**, *40*, 95–99. [CrossRef]

33. Blatov, V.A.; O'Keeffe, M.; Proserpio, D.M. Vertex-, face-, point-, Schläfli-, and Delaney-symbols in nets, polyhedra and tilings: Recommended terminology. *Cryst. Eng. Comm.* **2010**, *12*, 44–48. [CrossRef]

34. Krivovichev, S.V. Structural and topological complexity of zeolites: An information-theoretic analysis. *Microporous Mesoporous Mater.* **2013**, *171*, 223–229. [CrossRef]

35. Krivovichev, S.V. Which inorganic structures are the most complex? *Angew. Chem. Int. Ed.* **2014**, *53*, 654–661. [CrossRef]

36. Blatov, V.A.; Ilyushin, G.D.; Blatova, O.A.; Anurova, N.A.; Ivanov-Schits, A.K.; Dem'yanets, L.N. Analysis of migration paths in fast-ion conductors with Voronoi–Dirichlet partition. *Acta Crystallogr. Sect. B Struct. Sci.* **2006**, *62*, 1010–1018. [CrossRef]

37. Anurova, N.A.; Blatov, V.A.; Ilyushin, G.D.; Blatova, O.A.; Ivanov-Schits, A.K.; Dem'yanets, L.N. Migration maps of Li$^+$ cations in oxygen-containing compounds. *Solid State Ion.* **2008**, *179*, 2248–2254. [CrossRef]

38. Eremin, R.A.; Kabanova, N.A.; Morkhova, Y.A.; Golov, A.A.; Blatov, V.A. High-throughput search for potential potassium ion conductors: A combination of geometrical-topological and density functional theory approaches. *Solid State Ion.* **2018**, *326*, 188–199. [CrossRef]

39. Blatov, V.A.; Shevchenko, A.P.; Proserpio, D.M. Applied Topological Analysis of Crystal Structures with the Program Package ToposPro. *Cryst. Growth Des.* **2014**, *14*, 3576–3586. [CrossRef]

40. Perdew, J.P.; Burke, K.; Ernzerhof, M. Generalized Gradient Approximation Made Simple. *Phys. Rev. Lett.* **1996**, *77*, 3865–3868. [CrossRef] [PubMed]

41. Kresse, G.; Joubert, D. From ultrasoft pseudopotentials to the projector augmented-wave method. *Phys. Rev. B* **1999**, *59*, 1758–1775. [CrossRef]

42. Kresse, G.; Furthmüller, J. Vienna Ab-initio Simulation Package (VASP), V.5.4.4. Available online: www.vasp.at (accessed on 25 June 2021).

43. Monkhorst, H.J.; Pack, J.D. Special points for Brillouin-zone integrations. *Phys. Rev. B* **1976**, *13*, 5188–5192. [CrossRef]

44. Dornberger-Schiff, K. On the nomenclature of the 80 plane groups in three dimensions. *Acta Crystallogr.* **1959**, *12*, 173. [CrossRef]

45. Burns, P.C.; Grice, J.D.; Hawthorne, F.C. Borate minerals. I. Polyhedral clusters and foundamental building block. *Can. Mineral.* **1995**, *33*, 1131–1151.

46. Grell, H.; Dornberger-Schiff, K. Symbols for OD groupoid families referring to OD structures (polytypes) consisting of more than one kind of layer. *Acta Crystallogr. Sect. A* **1982**, *38*, 49–54. [CrossRef]

47. Ďurovič, S. Desymmetrization of OD Structures. *Krist. Tech.* **1979**, *14*, 1047–1053. [CrossRef]

48. Merlino, S. *EMU Notes in Mineralogy. Vol. 1. Modular Aspects of Minerals*; Merlino, S., Ed.; Eötvös University Press: Budapest, Hungary, 1997.

49. Rocha, J.; Lin, Z. Microporous Mixed Octahedral-Pentahedral-Tetrahedral Framework Silicates. *Rev. Mineral. Geochem.* **2005**, *57*, 173–201. [CrossRef]

50. Chukanov, N.V.; Pekov, I.V.; Rastsvetaeva, R.K. Crystal chemistry, properties and synthesis of microporous silicates containing transition elements. *Russ. Chem. Rev.* **2004**, *73*, 205–223. [CrossRef]

51. Chukanov, N.V.; Pekov, I.V. Heterosilicates with Tetrahedral-Octahedral Frameworks: Mineralogical and Crystal-Chemical Aspects. *Rev. Mineral. Geochem.* **2005**, *57*, 105–143. [CrossRef]

52. Baerlocher, C.; McCusker, L.B. Database of Zeolite Structures. Available online: http://www.iza-structure.org/databases/ (accessed on 25 June 2021).

53. Voronkov, A.A.; Ilyukhin, V.V.; Belov, N.V. Crystal chemistry of mixed frameworks—Principles of their formation. *Kristallografiya* **1975**, *20*, 556–566.

54. Sandomirskiy, P.A.; Belov, N.V. *Crystal Chemistry of Mixed Anionic Radicals*; Nauka: Moscow, Russia, 1984.

55. Ilyushin, G.D.; Blatov, V.A. Crystal chemistry of zirconosilicates and their analogs: Topological classification of MT frameworks and suprapolyhedral invariants. *Acta Crystallogr. Sect. B Struct. Sci.* **2002**, *58*, 198–218. [CrossRef]

56. Lesage, J.; Guesdon, A.; Raveau, B. RbGa$_3$ (P$_3$O$_{10}$)$_2$: A new gallium phosphate isotypic with RbAl$_3$(P$_3$O$_{10}$)$_2$. *Acta Crystallogr. Sect. C Cryst. Struct. Commun.* **2005**, *61*, i44–i46. [CrossRef]

57. Ruchkina, E.A.; Belokoneva, E.L. Structural features of lead iron borophosphates of alkali metals as analyzed in terms of topologically similar structural blocks. *Russ. J. Inorg. Chem.* **2003**, *48*, 1969–1978.

58. Topnikova, A.; Belokoneva, E.; Dimitrova, O.; Volkov, A.; Deyneko, D. Rb$_{1.66}$Cs$_{1.34}$Tb[Si$_{5.43}$Ge$_{0.57}$O$_{15}$]·H2O, a New Member of the OD-Family of Natural and Synthetic Layered Silicates: Topology-Symmetry Analysis and Structure Prediction. *Minerals* **2021**, *11*, 395. [CrossRef]

59. Belokoneva, E.L.; Reutova, O.V.; Dimitrova, O.V.; Volkov, A.S. Germanosilicate Cs$_2$In$_2$[(Si$_{2.1}$Ge$_{0.9}$)$_2$O$_{15}$](OH)$_2$ H$_2$O with a New Corrugated Tetrahedral Layer: Topological Symmetry-Based Prediction of Anionic Radicals. *Crystallogr. Rep.* **2020**, *65*, 566–572 [CrossRef]

minerals

Article

Twinning of Tetrahedrite—OD Approach

Emil Makovicky

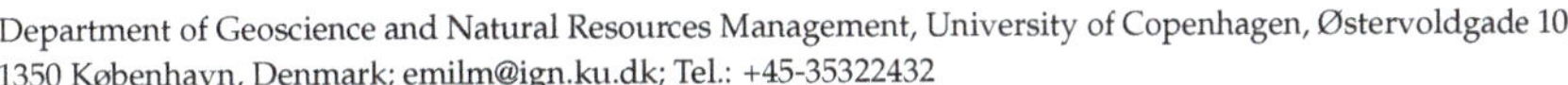

Department of Geoscience and Natural Resources Management, University of Copenhagen, Østervoldgade 10, 1350 København, Denmark; emilm@ign.ku.dk; Tel.: +45-35322432

Abstract: The common twinning of tetrahedrite and tennantite can be described as an order–disorder (OD) phenomenon. The unit OD layer is a one-tetrahedron-thick (111) layer composed of six-member rings of tetrahedra, with gaps between them filled with Sb(As) coordination pyramids and triangular-coordinated (Cu, Ag). The stacking sequence of six-member rings is ABCABC, which can also be expressed as a sequence of three consecutive tetrahedron configurations, named α, β, and γ. When the orientation of component tetrahedra is uniform, the α, β, γ, α sequence builds the familiar cage structure of tetrahedrite. However, when the tetrahedra of the β layer are rotated by 180° against those in the underlying α configurations and/or when a rotated α configuration follows after the β configuration (instead of γ), twinning is generated. If repeated, this could generate the ABAB sequence which would modify the structure considerably. If the rest of the structure grows as a regular cubic tetrahedrite structure, the single occurrence of the described defect sequences creates a twin.

Keywords: tetrahedrite; tennantite; twinning; order–disorder approach; tetrahedral framework

1. Introduction

check for **updates**

Citation: Makovicky, E. Twinning of Tetrahedrite—OD Approach. *Minerals* **2021**, *11*, 170. https://doi.org/10.3390/min11020170

Academic Editor: Giovanni Ferraris

Received: 19 January 2021
Accepted: 3 February 2021
Published: 7 February 2021

Publisher's Note: MDPI stays neutral with regard to jurisdictional claims in published maps and institutional affiliations.

Tetrahedrite is an old, long-known mineral species. It was known already to old miners as *fahlerz, weissgiltigerz, grey ore,* or *panabase,* under names mostly related to its macroscopic appearance in hand specimens. Its present name *"tetrahedrite"* was given by Haidinger [1] because of the common tetrahedral form shown by its crystals. The name *"tennantite"* was given to its As-based analogue, first described by W. and R. Phillips [2,3] from Cornwall. Early reports of the occurrence of Fe and Zn in tetrahedrite were the starting point of the long research path which, among other results, led to the chemical formula $Cu_{12}(Fe,Zn)_2(Sb,As)_4S_{13}$ for the most common tetrahedrite—tennantite solid solution. The voluminous literature concerned with the chemistry of natural and synthetic tetrahedrite and tennantite and with selected synonyms (e.g., *binnite* and *coppite*) has recently been summarized and referenced by Biagioni et al. [4]. The principal complication of the chemistry of this solid solution, the interplay of Fe^{3+} and Fe^{2+} in tetrahedrite and tennatite, has been studied by several authors (e.g., Makovicky et al. [5,6]; Andreasen et al. [7]; Nasonova et al. [8]). It does not alter the crystal structure principles of these minerals.

The crystal structure of tetrahedrite (Figure 1) was refined by Wuensch [9] using a sample from Horhausen, Westerwald (Germany). That of tennantite was refined by Wuensch et al. [10] starting with older data of Pauling and Neuman [11], and for the tennantite-(Cu) by [12]. Structures of silver varieties were refined by, e.g., Peterson and Miller [13], Johnson and Burnham [14], Rozhdestvenskaya et al. [15], and Welch et al. [16]. Karanović et al. [17] reported the crystal structure of mercurian tetrahedrite from Serbia, confirming the results of Kalbskopf [18]. Other crystal structure investigations on mercurian tetrahedrite were reported by Kaplunnik et al. [19], Foit and Hughes [20], as well as Biagioni et al. [21], and on hakite by Škácha et al. [22]. The crystal structure of synthetic Mn-tetrahedrite was described by Chetty et al. [23], whereas Barbier et al. [24] determined the crystal structure of a Ni-containing synthetic tetrahedrite. This count can be continued by recent studies of tetrahedrite by materials scientists. What is remarkable for the tetrahedrite–tennantite structure type, is the stability of structure motif (Figure 1) under all these element substitutions.

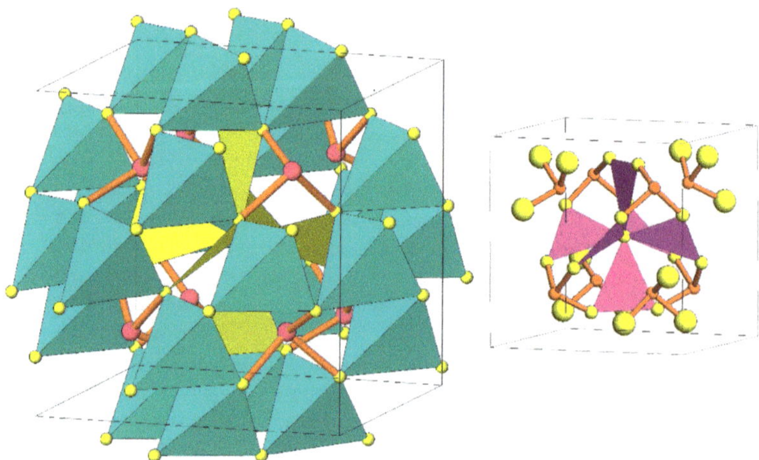

Figure 1. Crystal structure of tetrahedrite–tennantite. MeS$_4$ coordination tetrahedra (light blue) and majority of the (Cu,Ag)S$_3$ coordination triangles (yellow) are shown in polyhedral representation; the Sb(As)S$_3$ coordination pyramids as cation–anion bonds. Note filling of truncated-tetrahedron cavities in the tetrahedral framework by "spinners" composed of triangular co-ordinations. Inset "spinner" of (Cu,Ag)S$_3$ coordination triangles with (Sb,As)S$_3$ coordination pyramids surrounding spinner cavity.

2. Twinning of Tetrahedrite

The well-known twinning of tetrahedrite–tennantite (Figure 2) has been repeatedly described by (a selection of) those twin elements which constitute the difference between the holohedral cubic point-group symmetry $4/m\ \overline{3}\ 2/m$ and the point group symmetry of tetrahedrite, which is $\overline{4}3m$, as demonstrated by its morphology (Figure 2). In the present study, we attempt to describe the structural aspect of this twinning by means of defects which can occur during the growth of tetrahedrite crystals. We concentrate on the growth scheme of this structure, selecting the layer-by-layer mechanism, the most probable growth layers being one-tetrahedron-thick (111) in the cubic structure, which also are the principal crystal form of these minerals. Although the tetrahedrite structure often is "derived" from the sphalerite structure, presence of large cavities separated from one another by single-polyhedron-thick walls leads to a much more complicated configuration of (111) growth layers than found in sphalerite (Figure 3). Variation in stacking of these layers ought to be the reason for twinning. In the present model, we present a potentially free stacking variation of layers but within well-defined layer-match rules, i.e., we presume that they behave as order–disorder (OD) layers as defined by Dornberger-Schiff [25], Ďurovič [26], and Ferraris et al. [27], among others.

There is a limited number of types of such layers in any OD structure (only one layer type in our case), with own layer-group symmetry and 2D architecture. Their relationships in any layer pair (of two identical layers) is in the OD structure always described by the same set of symmetry operations. However, this does not guarantee fixed relationships, and periodicity, for triple layers and higher n-tuples, as it does in the majority of crystal structures. This results in a disordered layer stacking while maintaining the layer-to-layer relations for any pair of neighbors.

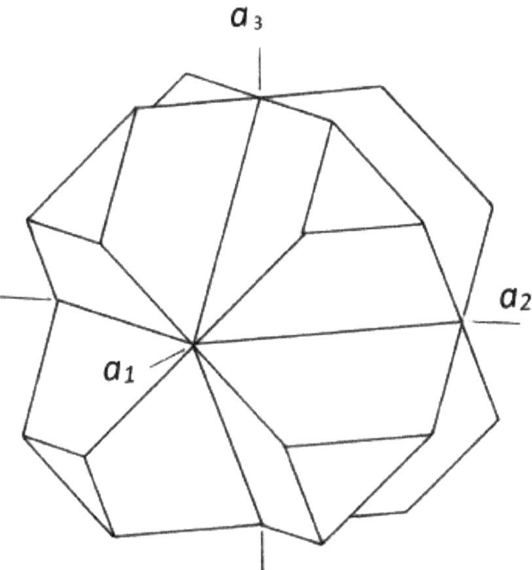

Figure 2. Interpenetration twin of tetrahedrite by reflection on $m\|(100)$. Twin symmetry elements are $m\|(001)$, $4\|[100]$, $2\|[110]$, and inversion center.

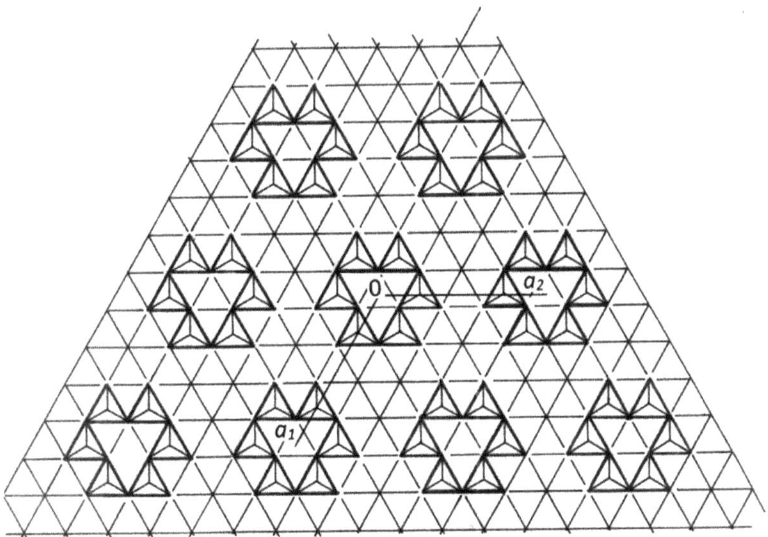

Figure 3. A single one-tetrahedron-thick (111) order–disorder OD layer of the tetrahedrite structure from Figure 1. Only the tetrahedral framework is shown. Tetrahedra present in the given layer are given in bold outlines, on a background of the net of virtual tetrahedron bases. Triangular openings of the six-member tetrahedron rings are filled (and their shape constrained) by $Sb(As)S_3$ coordination pyramids (not shown). The surrounding "empty" space of virtual tetrahedron bases represents different sections of the structure cavities which contain corresponding portions of the "spinners" with triangular Cu(Ag) coordinations (not drawn).

As a result of differing circumstances, mostly because of unfavorable distortion/modification of the ideal OD structure motif, instead of fully disordered layer sequences, frequent or even infrequent twinning can occur, representing faults in otherwise periodic layer sequences. In most cases, these can be described as twinning. In our case, it is the occurrence of Cu(Ag)$_6$S "spinners" [9] which fill the "collapsed sodalite-like cavities" [28] of the tetrahedron framework (Figure 1), that influences and defines the OD phenomena, because we expect these cavities to be strongly modified when a "faulty" layer sequence appears in the tetrahedrite-like structure.

3. The Concept of OD Layers

3.1. The Untwinned Layer Sequence

In tetrahedrite-tennantite, the OD layers are parallel to (111), and one tetrahedron thick. They are polar, with triangular bases of tetrahedra all oriented to one side (Figure 3), and the "free" tetrahedron vertices all turned towards the opposite side of the layer. The unit measure of all dimensions in the layer is the length of the edge of an *MeS*$_4$ coordination tetrahedron (which usually is occupied by Cu, Fe, Zn; Makovicky and Karup-Møller [29]); this results in a "sphalerite-like motif" of intermixed existing and virtual tetrahedra forming the layer, and determines the dimensions of the unit mesh (Figure 3). The layer is composed of isolated "collapsed" hexagonal rings of occupied tetrahedra, with three-fold symmetry (the OD layers do not have to be crystal-chemical layers and even can be disjoined although they must be periodic). The rings form vertices of a 2D hexagonal cell, with axes four tetrahedral edges long, and layer group 3*m*1 (Figure 3). This scheme is a pure tetrahedral OD scheme, in which we consider the cavities which contain Cu(Ag)$_6$S "spinners", and to some extent (e.g., concerning the orientation) also the (Sb,As) coordination pyramids, as a fill of the illustrated tetrahedral scheme. It should be stressed that all configurations observed in the OD layer are directly related to, and derived from, those observed in the 3D structure with cages and with (Sb,As) in inter-cage partitions (Figure 1).

The six-tetrahedra large (but after distortion three-fold) triangular openings of the spinner cage, which are kept constrained to three-fold symmetry by the (Sb,As)S$_3$ pyramids situated in the opening, will be called the α configurations (Figures 3 and 4). They were chosen as the origin of the trigonal 2D cell, which is 4 tetrahedron edges × 4 tetrahedron edges in size (Figure 3). In the adjacent portions of the (111) layer, on threefold axes, three pairs of such tetrahedra, from three surrounding α groups, are bonded via their base vertices with a Cu$_6$S-spinner [9] which is situated at (2/3, 1/3) of the 2D cell. This triangular area is the β-element of the planar pattern (Figures 3 and 4). Except via this spinner, the three α elements which surround the β-element (or "configuration") are not interconnected. The other triangular gap between three adjacent α configurations, which will be called "the γ-type", is surrounded and limited by horizontal edges of six tetrahedra from the α groups, which are situated at the origin and at two cell corners which have the $y = 1.0$ coordinate. This "antithesis" of the β configuration has tetrahedral sites at triangular corners vacant (Figures 3 and 4) and is situated at (1/3, 2/3). The outlined scheme of configurations is identical for all (111) layers of the {111} form.

The regular spinner-cage of untwinned cubic tetrahedrite is built by a sequence of configurations strung strictly along the line perpendicular to the (111) OD-layer: from bottom of Figure 4 upwards: α is covered by, and vertex-connected to, β; the latter in turn is vertex-connected to an overlying γ ring. After these larger β and γ configurations, the more constricted α configuration, parallel with the orientation and placement of the initial α, follows and closes the cage (Figures 3 and 4). (Sb, As) atoms are placed in the triangular cores of the α elements. One (Sb,As) atom is on the level of the α tetrahedra, three (Sb,As) atoms are on the β level, three on the γ level, and finally one is in the closing α ring (Figure 1). The (Sb,As) groups alternatively assume opposing orientations (in- and outward-oriented in respect to the cage).

As the α, β, and γ elements are parts of all OD layers, and the layers are identical, the outlined scheme means that, upwards, the OD layers in a layer sequence undergo shifts (defined here by the position of the α configuration) as follows: (0,0); (−1/3, −2/3); (−2/3, −1/3); (0,0). These shifts define the classical ABCABC stacking sequence of cubic close packing but with shift lengths forming a superstructure of the cubic close packing of a tetrahedra. These shifts preserve the same orientation of tetrahedra in all consecutive (111) OD layers (Figure 4). The described sequence yields the regular scheme of the undisturbed tetrahedrite/tennantite structure.

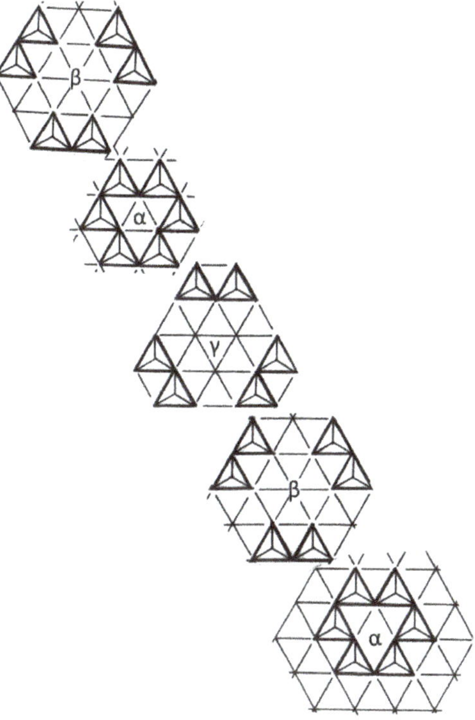

Figure 4. The truncated-tetrahedron cavity of the structural framework encompasses four consecutive OD layers (two of which are shared with the preceding and the following cavity). The corresponding configurations from one blown-out cavity are denoted as α, β, γ, α; then, the sequence repeats. Lateral shifts in the depicted sequence simulate those needed for vertex-fitting.

3.2. The Twinned Layer Sequence

The well-known twinning of tetrahedrite/tennantite requires that the tetrahedron orientation is altered by 180° rotation (Figure 2), together with the entire (111) OD layer, in comparison with the original sequence. Interconnection of tetrahedron vertices with the underlying layer must be preserved in this process (Figure 5).

If we examine the just growing 180°-rotated (111) layer of tetrahedra on the surface of tetrahedrite, the interconnection condition is satisfied when the 180°-rotated α element in the growing layer is anchored on vertices of the β-element in the starting layer (Figure 5, top). Then, in the adjacent portion of the growing rotated layer, the β-element adjacent to the said α element (both rotated in respect to such configurations in the starting layer) will be anchored on pointing vertices of the α element in the starting layer. The adjacent γ element in the rotated layer is, in a rotated fashion, anchored on the vertices of the γ element in the starting layer (Figure 5, right-hand corner). The α-on-β sequence is one of the mechanisms (structure defects) by which the twinned structure orientation is created.

In the resulting αβα′ sequence, the transfer of the first to the third layer represents a twofold screw axis running through the α rings, with a shift equal to the thickness of two layers and interlayer spaces. It is a trigonal antiprism of six corner tetrahedra, unlike the full tetrahedrite cage.

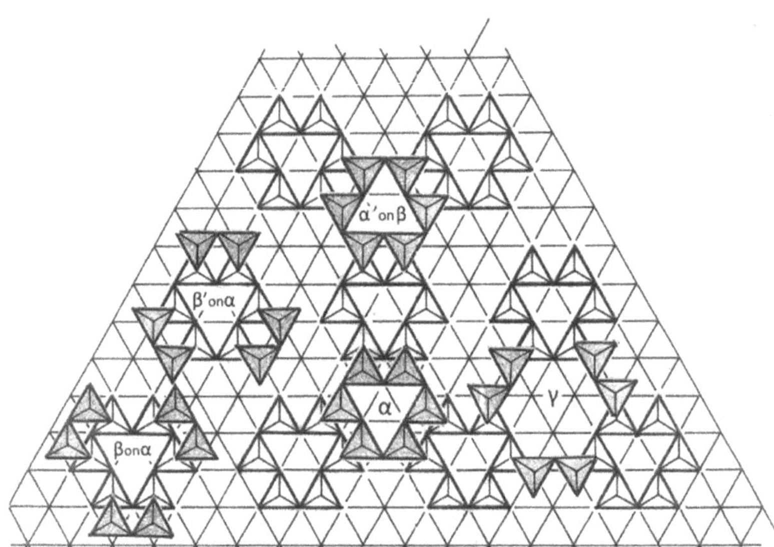

Figure 5. Tetrahedron-fit and vertex sharing of individual configurations from two (111) layers immediately following one another; the configurations are enumerated in the text. Configurations α, β, and γ for the unrotated layers are shown at the bottom of the figure, whereas stacking of α and β configurations for sequences which involve 180° rotation of consecutive layers are shown above them and are indicated by priming of the symbols.

The defect shift sequence can be defined as $(0, 0)$; $2_{(1\,1/3,\,2\,2/3,\,z)}$ and t$(-1\,1/3, -2\,2/3)$ i.e., rotation and displacement of the β element to the origin.

When centered on the α element in the starting layer, and attaching to its vertices the β element in the growing layer can assume two orientations (Figure 5, left). One of them has tetrahedra oriented in the fashion parallel to those in the preceding α element (and its entire starting (111) plane). After that, either a γ element in the next growing layer and then the usual tetrahedrite-like sequence can follow, or a rotated α element can follow after the attached β, as described in the preceding paragraph. In the other orientation, the β configuration in the growing layer is 180° rotated against that in the starting layer but it still fits with the vertices of the underlying α element. As the result of these two choices, the set of layer shifts either creates a normal tetrahedrite cage, αβγα, or it gives a modified sequence, αβ′γ′α′, or it even can generate the αβ′α sequence (which has been already mentioned). Presence of the rotated sequence means that the "(0,0) configuration' is followed by a layer with rotation and a shift $(-1\,1/3, -2\,2/3)$.

The just described αβ′ sequence produces a nicely interconnected openwork of tetrahedra as a basis of further growth. Both above outlined approaches, α′β and αβ′, give the same, identical result because placing the rotated α element on the β element automatically places the adjacent rotated β onto the α configuration. If repeated, instead of the ABCABC layer-stacking sequence which is typical for tetrahedrite, this sequence produces an ABABAB stacking of α rings, which are of the same polarity along the stacking axis, with all rings 180°-reversed (in a wurtzite-like fashion) in the B layers of the stacking formula. In this case, the gamma elements become open channels along the direction perpendicular to OD planes, unlike the cage-scheme observed in tetrahedrite.

3.3. Twin Symmetry in OD Description

The OD groupoid symbol, describing both the symmetry of individual OD layer and the symmetry operations transforming the *n*th layer into the (*n* + 1)th layer reads as

$$P \quad m \quad m \quad m \quad (3) \quad 1 \quad 1 \quad 1$$

$$\{ \quad n_{1/3,2} \quad n_{1/3,2} \quad n_{1/3,2} \quad (2_2) \quad c_2 \quad c_2 \quad c_2 \quad \}$$

where the individual elements relate to three horizontal crystallographic axes a_n, followed by the value for the c axis and directions parallel to it, and by three directions which halve the angles between adjacent a axes. Thus, the mirror planes present in the layer (Figure 6) are perpendicular to the a cell axes, and are interleaved by a full-unit-cell size gliding arrangement of α rings; these planes are extended into a layer pair as $n_{1/3,2}$ glide planes of the (11) orientation; these are active for the ABC sequence. The glide component of 1/3 is valid for all three periodic directions in which the α elements in $(n + 1)$ layer surround the initial element in the nth layer, although the periodicities in these three directions are not equal in their absolute length (in Å). The second subscript ('2') indicates a full OD layer-to-OD layer shift. Additional c-glide planes are oriented as (10) planes and are active only in the ABA sequence. Similar to the diagonal glide planes in the formula, the c glide planes all are c_2 operations [26], i.e., the layer-to-layer operations, and not the "classical" planes with a half-period translation component. They are placed alternatively between two more distant α rings and between two underlying and one overimposed α ring, with their more extensive overlap in projection (Figure 6). The latter sequence contains [001] 2_2 rotation axes. Thus, the ABC sequence is as follows.

$$P \quad m \quad m \quad m \quad (3) \quad 1 \quad 1 \quad 1$$

$$\{ \quad n_{1/3,2} \quad n_{1/3,2} \quad n_{1/3,2} \quad (1) \quad 1 \quad 1 \quad 1 \quad \}$$

and the ABA sequence is

$$P \quad m \quad m \quad m \quad (3) \quad 1 \quad 1 \quad 1$$

$$\{ \quad 1 \quad 1 \quad 1 \quad (2_2) \quad c_2 \quad c_2 \quad c_2 \quad \}$$

All the symmetry operations preserve the polar layer orientations. The layer-reversing symmetry operations are absent.

All other attempts of layer fitting result in small partial fits and huge misfits elsewhere in ring-like configurations. If the rest of the structure grows as a regular cubic tetrahedrite structure, the single occurrence of the described defect sequences creates a twin.

3.4. Penetration Twins

There are four equivalent planes in the {111} form, and each of them can give rise to the described twinning. What is the situation along [110], the meeting line of two such planes or even the [111] meeting point of three {111} planes? To answer the first question, when two α elements on the opposing {111} planes meet, they produce a characteristic group of four parallel tetrahedra (Figure 1). Closing of the space between such adjacent groups creates an undisturbed structure.

What is the situation in one growing (111) plane, which surrounds a patch of twin-oriented (111) plane in its middle? This situation is modeled in Figure 6, in which it can be seen that the rotated α configurations are displaced from their regular spacing in unrotated structure portions. This displacement can be described as separation by one row of virtual tetrahedral bases on all three equivalent {10} lattice planes in the trigonal plane-group (lattice). Displacement proceeds parallel to the axes of the trigonal plane lattice, along [01] in Figure 6, and the shift is by one edge length of the virtual triangular base of a tetrahedron. Each of the three displacement orientations can occur in two alternative directions (+ and − on the given axis) leading to potential order–disorder shifts between adjacent domains if more than one rotated structure patch appears in the growing layer. Propagation of the rotated structure patch to consecutive growth layers is assured by the differences in the position of the free tetrahedron vertices in the starting layer and in the 180°-rotated version of the structure.

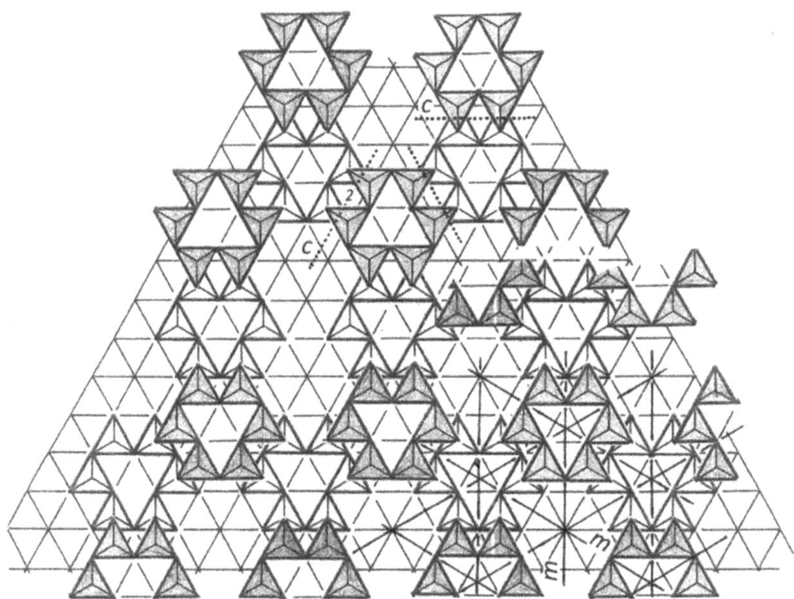

Figure 6. A two-layer sequence, with unrotated sequence of OD layers in the lower portion of the figure, and a 180°-rotated patch of the second (111) layer of tetrahedra in the upper portion of the figure. Incoherence between these two patches when they meet is accentuated in the right-hand central portions of the figure. Simplified symbols of symmetry operations indicate the reflection planes *m* in the individual OD layer, tied with the *n*-glide planes of the OD type (see full characterization in the text) for the unrotated layer pair, and the OD-type *c*-glide plane, and the corresponding 2_2 screw-axis, for the sequence with rotated OD layers.

3.5. (As,Sb)-Coordination Pyramids and (Cu,Ag)-Spinners

The OD phenomena of the purely tetrahedral OD scheme may be complicated by potential changes induced to the architecture of coordination pyramids and of lone electron pair schemes of As and Sb, and to those of $(Cu,Ag)_6S$ spinners in the cavities. Do we see configurational and compositional changes, and are there clusters of any kind which are forbidden?

Models reveal that the above described defect sequences allow three (Sb,As) pyramids on the level of β element in an arrangement as exists in the undisturbed tetrahedrite structure. Orientation of adjacent tetrahedra suggests that these pyramids are oriented outwards, out of the cage. The pyramid in the initial α ring points inwards, however, as do those in the γ element, whenever it follows. In the final α ring, the pyramid points out of the cage (or semi-cage) which we have just constructed, similar to the tetrahedra around it.

Whereas the question of (Sb,As) accommodation is surprisingly simple, that of spinners may be more complex. In the case of αβα′ sequence, only one half of a spinner (three arms) can fit in the reduced cavity. The central S atom of the original spinner may not fit this arrangement in the original form and the actual spinner re-arrangement is not known.

4. Conclusions

The model of twinning described here is based on order–disorder phenomena occurring when the tetrahedrite structure grows as a sequence of incremental (111) layers which are one tetrahedron thick. It does not require edge sharing of tetrahedra and resulting short cation–cation distances. As a chemical implication, it does not indicate excessively reducing conditions of formation.

The rough estimate of the local compositional problems for twinned layer configuration is difficult to give because changes in spinner configurations on twinning are not known. However, from the initial discussion (above), it follows that all OD layers have identical composition, notwithstanding their shifts, and the disordered structure should have the same composition as the ordered ABCABC structure. This indicates that the usually observed formation conditions for tetrahedrite–tennantite should not influence the presence or frequency of its twinning, although the presence of more exotic substitutes (with unusual atom radii) might do so.

Funding: This research received no external funding.

Data Availability Statement: Crystallographic data on tetrahedrite are publicly available.

Conflicts of Interest: The author declares no conflict of interest.

References

1. Haidinger, W. *Handbuch der Bestimmenden Mineralogie*; Braumüller and Seidel: Wien, Austria, 1845; pp. 563–570.
2. Phillips, R. Analysis of the copper ore, described in the preceding paper. *Q. J. Sci. Lit. Arts* **1819**, *7*, 100–102.
3. Phillips, W. Description of an ore of copper from Cornwall. *Q. J. Sci. Lit. Arts* **1819**, *7*, 95–100.
4. Biagioni, C.; George, L.L.; Cook, N.J.; Makovicky, E.; Moëlo, Y.; Pasero, M.; Sejkora, J.; Stanley, C.J.; Welch, M.D.; Bosi, F. The tetrahedrite group: Nomenclature and classification. *Am. Mineral.* **2020**, *105*, 109–122. [CrossRef]
5. Makovicky, E.; Forcher, K.; Lottermoser, W.; Amthauer, G. The role of Fe^{2+} and Fe^{3+} in synthetic Fe-substituted tetrahedrite. *Mineral. Petrol.* **1990**, *43*, 73–81. [CrossRef]
6. Makovicky, E.; Tippelt, G.; Forcher, K.; Lottermoser, W.; Karup-Møller, S.; Amthauer, G. Mössbauer study of Fe-bearing synthetic tennantite. *Can. Mineral.* **2003**, *41*, 1125–1134. [CrossRef]
7. Andreasen, J.W.; Makovicky, E.; Lebech, B.; Karup Møller, S. The role of iron in tetrahedrite and tennantite determined by Rietveld refinement of neutron powder diffraction data. *Phys. Chem. Miner.* **2008**, *35*, 447–454. [CrossRef]
8. Nasonova, D.I.; Presniakov, I.A.; Sobolev, A.V.; Verchenko, V.Y.; Tsirlin, A.A.; Wei, Z.; Dikarev, E.; Shevelkov, A.V. Role of iron in synthetitc tetrahedrites revisited. *J. Solid State Chem.* **2016**, *235*, 28–35. [CrossRef]
9. Wuensch, B.J. The crystal structure of tetrahedrite, $Cu_{12}Sb_4S_{13}$. *Z. Krist.* **1964**, *119*, 437–453. [CrossRef]
10. Wuensch, B.J.; Takéuchi, Y.; Nowacki, W. Refinement of the crystal structure of binnite, $Cu_{12}As_4S_{13}$. *Z. Krist.* **1966**, *123*, 1–20. [CrossRef]
11. Pauling, L.; Neuman, E.W. The crystal structure of binnite $(Cu,Fe)_{12}As_4S_{13}$ and the chemical composition and structure of minerals of the tetrahedrite group. *Z. Krist.* **1934**, *88*, 54–62. [CrossRef]
12. Makovicky, E.; Karanović, L.; Poleti, D.; Balić-Žunić, T.; Paar, W.H. Crystal structure of copper-rich unsubstituted tennantite, $Cu_{12.5}As_4S_{13}$. *Can. Mineral.* **2005**, *43*, 679–688. [CrossRef]
13. Peterson, R.C.; Miller, I. Crystal structure and cation distribution in freibergite and tetrahedrite. *Mineral. Mag.* **1986**, *50*, 717–721. [CrossRef]
14. Johnson, M.L.; Burnham, C.W. Crystal structure refinement of an arsenic-bearing argentian tetrahedrite. *Am. Mineral.* **1985**, *70*, 165–170.
15. Rozhdestvenskaya, I.V.; Zayakina, N.V.; Samusikov, V.P. Crystal structure features of minerals from a series of tetrahedrite-freibergite. *Mineral. Zhurnal* **1993**, *15*, 9–17. (In Russian)
16. Welch, M.D.; Stanley, C.J.; Spratt, J.; Mills, S.J. Rozhdestvenskayaite $Ag_{10}Zn_2Sb_4S_{13}$ and argentotetrahedrite $Ag_6Cu_4(Fe^{2+}, Zn)_2Sb_4S_{13}$: Two Ag-dominant members of the tetrahedrite group. *Eur. J. Mineral.* **2018**, *30*, 1163–1172. [CrossRef]
17. Karanović, L.; Cvetković, L.; Poleti, D.; Balić-Žunić, T.; Makovicky, E. Structural and optical properties of schwazite from Dragodol (Serbia). *Neues Jahrb. Mineral. Mon.* **2003**, 503–520. [CrossRef]
18. Kalbskopf, R. Strukturverfeinerung des Freibergits. *Tschermaks Mineral. Petrogr. Mitt.* **1972**, *18*, 147–155. [CrossRef]
19. Kaplunnik, L.N.; Pobedimskaya, E.A.; Belov, N.V. The crystal structure of schwazite $(Cu_{4.4}Hg_{1.6})Cu_6Sb_4S_{12}$. *Dokl. Akad. Nauk SSSR* **1980**, *253*, 105–107.
20. Foit, F.F.; Hughes, J.M. Structural variations in mercurian tetrahedrite. *Am. Mineral.* **2004**, *89*, 159–163. [CrossRef]
21. Biagioni, C.; Sejkora, J.; Musetti, S.; Velebil, D.; Pasero, M. Tetrahedrite-(Hg), a new 'old' member of the tetrahedrite group. *Mineral. Mag.* **2020**, *84*, 584–592. [CrossRef]
22. Škácha, P.; Sejkora, J.; Palatinus, L.; Makovicky, E.; Plášil, J.; Macek, I.; Goliáš, V. Hakite from Příbram, Czech Republic: Compositional variability, crystal structure and the role in Se mineralization. *Mineral. Mag.* **2016**, *80*, 1115–1128. [CrossRef]
23. Chetty, R.; Prem Kumar, D.S.; Rogl, G.; Rogl, P.; Bauer, E.; Michor, H.; Suwas, S.; Puchegger, S.; Giester, G.; Mallik, R.C. Thermoelectric properties of a Mn substituted synthetic tetrahedrite. *Phys. Chem. Chem. Phys.* **2015**, *17*, 1716–1727. [CrossRef]

24. Barbier, T.; Lemoine, P.; Gascoin, S.; Lebedev, O.I.; Kaltzoglou, A.; Vaquiero, P.; Powell, A.V.; Smith, R.I.; Guilmeau, E. Structural stability of the synthetic thermoelectric ternary and nickel-substituted tetrahedrite phases. *J. Alloys Compd.* **2015**, *634*, 253–262. [CrossRef]
25. Dornberger-Schiff, K. *Lehrgang über OD-Strukturen*; Akademie-Verlag: Berlin, Germany, 1966; 135p.
26. Ďurovič, S. Fundamentals of the OD theory. *EMU Notes Mineral.* **1997**, *1*, 3–28.
27. Ferraris, G.; Makovicky, E.; Merlino, S. *Crystallography of Modular Materials*; Oxford University Press: Oxford, UK, 2004; 370p.
28. Makovicky, E. Crystal structures of sulfides and other chalcogenides. *Rev. Mineral. Geochem.* **2006**, *61*, 7–125. [CrossRef]
29. Makovicky, E.; Karup-Møller, S. Exploratory studies on substitution of minor elements in synthetic tetrahedrite. Part I. Substitution by Fe, Zn, Co, Ni, Mn, Cr, V and Pb. Unit-cell parameter changes on substitution and the structural role of "Cu^{2+}" *Neues Jahrb. Mineral. Abh.* **1994**, *167*, 89–123.

Article

Rb$_{1.66}$Cs$_{1.34}$Tb[Si$_{5.43}$Ge$_{0.57}$O$_{15}$]·H$_2$O, a New Member of the OD-Family of Natural and Synthetic Layered Silicates: Topology-Symmetry Analysis and Structure Prediction

Anastasiia Topnikova [1], Elena Belokoneva [1,*], Olga Dimitrova [1], Anatoly Volkov [1] and Dina Deyneko [2]

[1] Geological Faculty, M.V. Lomonosov Moscow State University, Leninskie Gory 1, 119991 Moscow, Russia; nastya_zorina@rambler.ru (A.T.) dimitrova@list.ru (O.D.); toljha@yandex.ru (A.V.)

[2] Chemical Faculty, M.V. Lomonosov Moscow State University, Leninskie Gory 1, 119991 Moscow, Russia; deynekomsu@gmail.com

* Correspondence: elbel@geol.msu.ru

Abstract: Crystals of new silicate-germanate Rb$_{1.66}$Cs$_{1.34}$Tb[Si$_{5.43}$Ge$_{0.57}$O$_{15}$]·H$_2$O have been synthesized hydrothermally in a multi-component system TbCl$_3$:GeO$_2$:SiO$_2$ = 1:1:5 at T = 280 °C and P = 100 atm. K$_2$CO$_3$, Rb$_2$CO$_3$ and Cs$_2$CO$_3$ were added to the solution as mineralizers. The crystal structure was solved using single crystal X-ray data: a = 15.9429(3), b = 14.8407(3), c = 7.2781(1) Å, sp. gr. *Pbam*. New Rb,Cs,Tb-silicate-germanate consists of a [Si$_{5.43}$Ge$_{0.57}$O$_{15}$]$_{\infty\infty}$ corrugated tetrahedral layer combined by isolated TbO$_6$ octahedra into the mixed microporous framework as in synthetic K$_3$Nd[Si$_6$O$_{15}$]·2H$_2$O, K$_3$Nd[Si$_6$O$_{15}$] and K$_3$Eu[Si$_6$O$_{15}$]·2H$_2$O with the cavities occupied by Cs, Rb atoms and water molecules. Luminescence spectrum on new crystals was obtained and analysed. A comparison with the other representatives of related layered natural and synthetic silicates was carried out based on the topology-symmetry analysis by the OD (order-disorder) approach. The wollastonite chain was selected as the initial structural unit. Three symmetrical ways of forming ribbon from such a chain and three ways of further connecting ribbons to each other into the layer were revealed and described with symmetry groupoids. Hypothetical structural variants of the layers and ribbons in this family were predicted.

Keywords: RE-silicate-germanate; hydrothermal synthesis; layered silicates; modular approach; wollastonite chain; topology-symmetry analysis; OD theory; structure prediction; luminescence properties

check for
updates

Citation: Topnikova, A.; Belokoneva, E.; Dimitrova, O.; Volkov, A.; Deyneko, D. Rb$_{1.66}$Cs$_{1.34}$Tb[Si$_{5.43}$Ge$_{0.57}$O$_{15}$]·H$_2$O, a New Member of the OD-Family of Natural and Synthetic Layered Silicates: Topology-Symmetry Analysis and Structure Prediction. *Minerals* **2021**, *11*, 395. https://doi.org/10.3390/min11040395

Academic Editors: Mariko Nagashima and Giovanni Ferraris

Received: 15 February 2021
Accepted: 7 April 2021
Published: 9 April 2021

Publisher's Note: MDPI stays neutral with regard to jurisdictional claims in published maps and institutional affiliations.

1. Introduction

Si and Ge elements have equal formal charge 4+ and tetrahedral coordination. In the crystal structures, Si and Ge occur together only as the isomorphic substitution Si-Ge in tetrahedra. Both elements are present isomorphically in the minerals sanbornite, milarite, albite, perrierite, farmakosiderite, garnet, titanite, and zeolite analogues [1]. Lone pair heavy metals act together with different anionic units containing mixing components. Such an approach is actually used in materials design to result in promising properties. There are no silicate-germanates with Tb in the nature or synthetic compounds according to [1,2]. However, there are a lot of original synthetic Tb-silicates with different anionic radicals [2,3]: ortho- NaTb$_9$(SiO$_4$)$_6$O$_2$ [4], KTb$_9$(SiO$_4$)$_6$O$_2$ and Cd$_2$Tb$_8$(SiO$_4$)$_6$O$_2$ [5] (structural type of apatite), Na$_5$Tb$_4$(OH)[SiO$_4$]$_4$ [6]; diortho- Tb$_2$Si$_2$O$_7$ [7], K$_3$TbSi$_2$O$_7$ [8], Tb$_4$S$_3$Si$_2$O$_7$ [9]; triortho- K$_3$TbSi$_3$O$_8$(OH)$_2$ [10]; tetraortho-groups Ba$_2$Tb$_2$Si$_4$O$_{13}$ [11]; six-membered rings Na$_3$Tb$_3$Si$_6$O$_{18}$·H$_2$O (synthetic gerenite) [12]; chains—unbranched Rb$_2$TbGaSi$_4$O$_{12}$ [13], wollastonite Na$_2$Tb$_{1.08}$Ca$_{2.92}$Si$_6$O$_{18}$H$_{0.8}$ [14] and spiral Na$_3$TbSi$_3$O$_9$·3H$_2$O [15]; layers in Cs$_3$TbSi$_4$O$_{10}$F$_2$ [16] and K$_{7.04}$Tb$_3$Si$_{12}$O$_{32.02}$·1.36H$_2$O [17]; sheets in Cs$_3$TbSi$_8$O$_{19}$·2H$_2$O [18] and Na$_4$K$_2$Tb$_2$Si$_{16}$O$_{38}$·10H$_2$O [19]. For some of these compounds, the luminescence properties were investigated [6,10,13–15,18,19].

Silicate crystal structures with the layers are presented [20,21] as a result of condensation of various chains. The formation of chains, layers, or frameworks by linking tetrahedra by the symmetry elements are partially addressed in a monograph [20], where the chains of Ba-silicates and symmetry elements responsible for their formation are identified. Important aspects of the modular approach and OD description of crystal structures were analyzed in [22]. For careful analysis of the similarities and differences in crystal structures, it is necessary to separate building units or modules which may be similar in different minerals, for example, layers in micas. The use of symmetry, a fundamental concept in crystallography, is a key tool for the description of structural families and for the construction of anionic radicals. Such an approach was suggested in [23] for layered ordered crystal structures in which a significant disorder component may be presented (OD theory). The local symmetry of layers or rods leads to structural variants of their conjugation and allows to predict new crystal structures. In the OD family, sursassite-pumpellyite-ardennite, the fourth member, was predicted and confirmed by high-resolution electron microscopy [24]. Strict symmetric rules dictate all possibilities in real or hypothetical crystal structures. The symmetry approach, based on the principles of OD theory [23], was developed for the borates [25] at all levels of condensation from the initial isolated tetrahedron to the chain, layer, and framework anionic units and described by groupoids of different ranks. As mentioned in the investigation of the crystal structure of the chain diborate $GdH[B_2O_5]$ [26], the results are the same for borates, silicates, and other tetrahedral radicals. Thus, the borate chain in vimsite $Ca[B_2O_2(OH)_4]$ is identical to the pyroxene chain. The formation of ribbons, layers, and frameworks in the well-known silicates are considered in [26]. The letters U (upward) and D (downward), which are commonly used in the literature for the description of tetrahedral anionic groups (chains, layers, and frameworks) actually reflects the absence or presence of inversion symmetry elements [27]. Two-chain ribbons are present in palygorskite $Mg_5[Si_4O_{10}]_2(OH)_2 \cdot 8H_2O$, and three-chain ribbons in sepiolite $Mg_4[Si_6O_{15}](OH)_2 \cdot 6H_2O$. In both minerals, pyroxene chains are related by the symmetry elements m_y, -1, m_y, $-1 \dots$ (palygorskite) or m_y, m_y, 2_x, m_y, m_y, $2_x \dots$ (sepiolite) [26]. The crystal structure of α-celsian $Ba[Al_2Si_2O_8]$ demonstrates next step of condensation: nonpolar double-decker sheets are formed of polar mica-like layers multiplied by the mirror plane m_z via sharing of the apical vertices of the tetrahedra.

Synthesis of a new Rb,Cs,Tb-silicate-germanate, its structure solution and crystal chemical comparison with known related natural and synthetic silicates led us to using topology-symmetry analysis of OD theory. The following rubricating of known and predicted layered crystal structures is presented in this work. Luminescence properties are also characterized.

2. Materials and Methods

2.1. Synthesis of Crystals

The crystals of a new Rb,Cs,Tb-silicate-germanate were synthesized by a hydrothermal method in the system, containing oxide and chloride components in the mass ratio $TbCl_3:GeO_2:SiO_2 = 1:1:5$ that corresponds to 1.0 g (0.003 mol) $TbCl_3$, 1.0 g (0.001 mol) GeO_2 and 5.0 g (0.017 mol) SiO_2. All the reagents were of analytical grade. The mass ratio of solid and liquid phases was 1:5. K_2CO_3, Rb_2CO_3 and Cs_2CO_3 were added at the solution as mineralizers. The phase formation occurred at a pH 3 (measured after the completion of the reaction). The synthesis was carried out at the temperature of 280 °C and pressure of 100 atm. A standard autoclave (capacity 5 to 6 cm^3) lined with Teflon was used. The characteristics of experiment were limited by the kinetics of the hydrothermal reactions and the instrumental capabilities. The reaction went to completion during heating for 18 to 20 days; followed by cooling to room temperature for over 24 h. The grown crystals were isolated by filtering the stock solution, washed with water and finally dried at room temperature. The small colorless transparent prismatic crystals, splices and brushes of crystals were found in the reaction products. The yield of the crystals was about 50 vol.%. Determination of the unit cell parameters on single crystals was performed using pre-experiment on Xcalibur S diffractometer (CCD area detector; graphite-monochromated Mo-K$_\alpha$ radiation). The chemical composition was determined using a Jeol JSM-6480LV scanning electronic microscope combined with WDX analysis (Jeol, Osaka, Japan). The qualitative test revealed the presence of Tb, Cs, Rb, Si, Ge and O.

2.2. Luminescence Study

Photoluminescence emission (PL) and excitation (PLE) spectra were recorded on an Agilent Cary Eclipse fluorescence spectrometer (Agilent Technology, Malaysia) with a 75 kW xenon light source (pulse length τ = 2 µs, pulse frequency ν = 80 Hz, wavelength resolution 0.5 nm; PMT Hamamatsu R928). For correct determination of photoluminescent properties, the measurements were performed on three portions of the crystals. The photoluminescence spectra of all samples were obtained under similar experimental conditions to compare the relative emission intensities and reduce the error. The experiment showed complete reproduction of the photoluminescence data for these three portions. The obtained spectra were corrected for the sensitivity of the spectrometer.

In Figure 1a, the PLE spectrum of $Rb_{1.66}Cs_{1.34}Tb[Si_{5.43}Ge_{0.57}O_{15}]\cdot H_2O$ is shown. According to the results, the Tb^{3+} ion can be excited in different ways: within intraconfigural $4f^8$ transitions, and by interconfigural $4f^8-4f^75d^1$ transition [28]. Several bands at 300 to 500 nm are attributed to f-f transitions of Tb^{3+} ions from the $^7F_6$ ground state to the $^5H_6$ (303 nm), $^5H_7$ (320 nm), $^5L_9$ (360 nm), $^5D_3$ (378 nm) and $^5D_4$ (480 nm). The most intensive $^5D_3$ transition was not split.

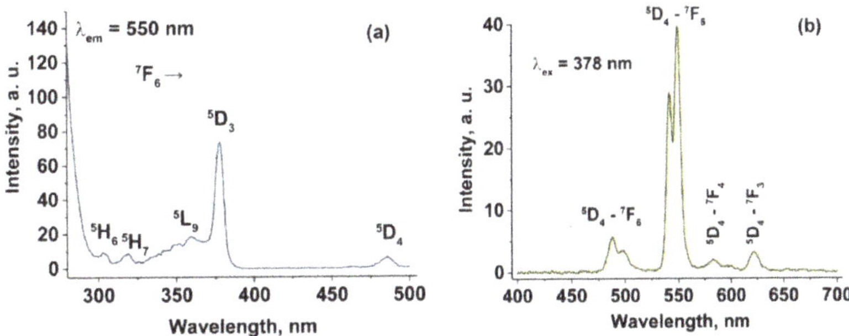

Figure 1. $Rb_{1.66}Cs_{1.34}Tb[Si_{5.43}Ge_{0.57}O_{15}]\cdot H_2O$: PLE spectrum (**a**) and PL spectrum at λ_{exc} = 378 nm (**b**).

In Figure 1b, the PL spectrum of $Rb_{1.66}Cs_{1.34}Tb[Si_{5.43}Ge_{0.57}O_{15}]\cdot H_2O$ controlled at λ_{exc} = 378 nm is shown. The spectrum consists of $^5D_4 \rightarrow {}^7F_J$ (J = 3–6) optical transitions of Tb^{3+} ion. The emission from the hypersensitive $^5D_4-{}^7F_5$ transition at 540 nm is predominant. There was no emission from the higher lying $^5D_3$ level to ^7F_J states (Figure 1b). Generally, the absence of these transitions is due to the presence of efficient cross-relaxation processes [29]. The deactivation in the $^5D_3$ emitting state such as $^5D_3 \rightarrow {}^5D_4$ and $^7F_6 \rightarrow {}^7F_0$ or $^5D_3 \rightarrow {}^7F_0$ and $^7F_6 \rightarrow {}^5D_4$ [30,31] is observed since the concentration of Tb^{3+} is relatively high. The main peak is split into two Stark components. This, apparently, is due to the presence of two non-equivalent positions occupied by Tb^{3+} in the crystal structure. Only the $^5D_4-{}^7F_J$ (J = 3–6) transition lines have a measurable intensity. It was noted that the luminescence intensity of the $^5D_4 \rightarrow {}^7F_3$ transition could become comparable to that of the main green $^5D_4 \rightarrow {}^7F_5$ transition of Tb^{3+}, when the crystal field is strong [32]. In $Rb_{1.66}Cs_{1.34}Tb[Si_{5.43}Ge_{0.57}O_{15}]\cdot H_2O$ the $^5D_4 \rightarrow {}^7F_3$ compared to $^5D_4 \rightarrow {}^7F_5$ is less, by approximately 10 times. So, the studied crystal has a low crystal field strength. In addition, the ratio between the $^5D_4 \rightarrow {}^7F_5$ transition (green band at 490 nm) to the $^5D_4 \rightarrow {}^7F_6$ transition (blue band at 540 nm) is known as green-to-blue fluorescence factor (G/B) [33]. The G/B (Tb^{3+}) determines the asymmetry of the local environment around the Tb^{3+} ions and character bonding (covalent/ionic) between Tb^{3+} and O^{2-}. The G/B factor for Tb^{3+}

ions in $Rb_{1.66}Cs_{1.34}Tb[Si_{5.43}Ge_{0.57}O_{15}]\cdot H_2O$ is 5.01 has been calculated. The obtained value is rather high, which suggests more covalent character of the bonding between terbium and oxygen ions. This is due to the presence of heavy rubidium and cesium atoms in the crystal composition.

3. Results

Structure Solving and Description

A small colorless transparent short-prismatic crystal with a size of $0.10 \times 0.05 \times 0.04$ mm was selected for the single-crystal X-ray study. The diffraction experiment was carried out on an Xcalibur-S diffractometer (Oxford Diffraction, Oxford, UK) with a graphite monochromatized Mo-K_α radiation source ($\lambda = 0.71073$ Å) and a CCD detector (ω scanning mode). The data were integrated using the CrysAlis Pro Agilent Technologies (v.1.1713735 2014) software [34] and corrected for the Lorentz and polarization factors. The refined orthorhombic unit-cell parameters are $a = 15.9429(3)$, $b = 14.8407(3)$, $c = 7.2781(1)$ Å $V = 1722.03(6)$ Å^3. A structural model was found by the direct method determination using SHELXS [35] within the WinGX v2018.3 [36] software in the suggested space group *Pbam* in agreement with the systematically absent reflections. At the first step, Tb1, Tb2, Rb1, Rb2, Cs1, Si and several of O sites were found. The remaining O sites were detected in different Fourier syntheses and were introduced in the model. As the temperature displacement parameters for all Si sites and Rb2 site were decreased, a small amount of Ge was isomorphically added in the Si sites (($Si_{0.93}Ge_{0.07}$)1, ($Si_{0.88}Ge_{0.12}$)2, ($Si_{0.91}Ge_{0.09}$)3, ($Si_{0.92}Ge_{0.08}$)4) and Cs—in Rb2 site ($Rb_{0.66}Cs_{0.34}$)2 which significantly improved the R-factor. The occupations of these sites were found first by changing them step by step in a search for the minimum R-factor with the control of the temperature displacement parameters. After that, we applied the procedure of the refinement of tetrahedral position occupations and (Rs,Cs)2 position occupations, which made it possible to improve the result. The O11 atom was identified as the oxygen of a water molecule based on Pauling's bond valence distribution [37] (Table 1). The resulting chemical formula is $Rb_{1.66}Cs_{1.34}Tb[Si_{5.43}Ge_{0.57}O_{15}]\cdot H_2O$ $Z = 4$. The structural model in a space group *Pbam* was refined using the least squares procedure in anisotropic approximation of the atomic displacements and with the refinement of the weighting scheme using SHELXL [38]. The absorption of crystal was not corrected because it was negligible $\mu r_{max} = 0.63$ and did not influence the result. Crystallographic data, atomic coordinates and selected bonds are presented in Tables 2–4. CCDC CSD 2,062,486 contains crystallographic data for this paper. These data can be obtained free of charge via www.ccdc.cam.ac.uk/data_request/cif (accessed on 24 March 2021). Illustrations were produced using ATOMS (v. 5.1) [39] and CORELDRAW (v. 21.0.0.593, 2019) programs.

Table 1. Pauling's balance of valences for $Rb_{1.66}Cs_{1.34}Tb[Si_{5.43}Ge_{0.57}O_{15}]\cdot H_2O$.

	$Tb1^{3+}$ C.N. = 6 0.25*	$Tb2^{3+}$ C.N. = 6 0.25*	$Cs1^+$ C.N. = 11 0.5*	$Rb1^+$ C.N. = 9 0.5*	$(Rb,Cs2)^+$ C.N. = 7 0.5*	$T1^{4+}$ C.N. = 4 0.5*	$T2^{4+}$ C.N. = 4 1.0*	$T3^{4+}$ C.N. = 4 1.0*	$T4^{4+}$ C.N. = 4 0.5*	Σexp	$\Sigma theor$
$O1^{2-}$ 0.5 *			0.045	0.056				1.0		−1.1	−1.0
$O2^{2-}$ 0.5 *					0.071	0.5			0.5	−1.071	−1.0
$O3^{2-}$ 1.0 *		0.125 × 4	0.045 × 4		0.071 × 2			1.0		−1.825	−2.0
$O4^{2-}$ 1.0 *	0.125 × 4			0.056 × 4			1.0			−1.722	−2.0
$O5^{2-}$ 1.0 *				0.056 × 2		0.5 × 2		1.0		−2.111	−2.0
$O6^{2-}$ 1.0 *					0.071 × 2		1.0	1.0		−2.143	−2.0
$O7^{2-}$ 0.5 *		0.125 × 2	0.045 × 2		0.071				0.5	−0.912	−1.0
$O8^{2-}$ 1.0 *			0.045 × 2				1.0		0.5 × 2	−2.091	−2.0
$O9^{2-}$ 0.5 *	0.125 × 2				0.071	0.5				−0.821	−1.0
$O10^{2-}$ 0.5 *			0.045	0.056				1.0		−1.1	−1.0
$O11w^{2-}$ 0.5 *			0.045	0.056						−0.1	0
Σ	+0.75	+0.75	+0.5	+0.5	+0.5	+2	+4	+4	+2	15	15

* These values correspond to multiplicities scaled to common position multiplicity equal to 1.0.

Table 2. Crystal data and structure refinement for $Rb_{1.66}Cs_{1.34}Tb[Si_{5.43}Ge_{0.57}O_{15}]\cdot H_2O$.

Formula	$Rb_{1.66}Cs_{1.34}Tb[Si_{5.43}Ge_{0.57}O_{15}]\cdot H_2O$
formula weight (g/mol)	930.48
T (K)	293(2)
crystal system	Orthorhombic
space group, Z	*Pbam*, 4
a (Å)	15.9429(3)
b (Å)	14.8407(3)
c (Å)	7.2781(1)
V (Å^3)	1722.03(6)
crystal size (mm)	$0.10 \times 0.05 \times 0.04$
ρ_{calc} (g/cm^3)	3.532
μ (mm^{-1})	12.610
F(000)	1677
wavelength (Å)	0.71073
θ range/deg.	2.75–30.78
limiting indices	$-22 \leq h \leq 22, -21 \leq k \leq 20, -10 \leq l \leq 10$
refl. collected/unique	28316/2756 [R_{int} = 0.0695]
completeness to theta	99.9
data/restraints/parameters	2756/0/143
GOF	1.187
R_1, wR_2 [1] $[I > 2\sigma(I)]$	0.0541, 0.0885
R_1, wR_2 (all data) [1]	0.0670, 0.0926
$\Delta\rho_{max}$ and $\Delta\rho_{min}$ (e Å^{-3})	1.726 and -2.318

[1] $R(F) = \sum ||F_o| - |F_c||/\sum |F_o|$ and $wR_2 = [\sum w(F_o^2 - F_c^2)^2/\sum w(F_o^2)^2]^{1/2}$ for $F_o^2 > 2\sigma(F_o^2)$.

Table 3. Atomic coordinates and atomic displacement parameters (U, Å^2) for $Rb_{1.66}Cs_{1.34}Tb[Si_{5.43}Ge_{0.57}O_{15}]\cdot H_2O$. U_{eq} is defined as one third of the trace of the orthogonalized U_{ij} tensor.

Atoms	Wyckoff Position, Point Symm.	S.o.f.	X	Y	Z	U_{eq}
Cs1	4*g*, *m*	1.0	0.1306(1)	0.5459(1)	0	0.0255(2)
Rb1	4*h*, *m*	1.0	0.2830(1)	0.8920(1)	0.5	0.0408(4)
(Rb, Cs)2	4*g*, *m*	0.66, 0.34	0.4222(1)	0.4033(1)	0	0.0250(2)
Tb1	2*b*, 2/*m*	1.0	0.5	0.5	0.5	0.00832(14)
Tb2	2*d*, 2/*m*	1.0	0	0.5	0.5	0.00750(14)
(Si, Ge)1	4*h*, *m*	0.93, 0.07	0.0874(1)	0.7829(2)	0.5	0.0083(7)
(Si, Ge)2	8*i*, 1	0.88, 0.12	0.3535(1)	0.6345(1)	0.7838(2)	0.0077(5)
(Si, Ge)3	8*i*, 1	0.91, 0.09	0.0395(1)	0.3024(1)	0.2181(2)	0.0087(5)
(Si, Ge)4	4*h*, *m*	0.92, 0.08	0.2068(1)	0.6291(2)	0.5	0.0060(7)
O(1)	4*g*, *m*	1.0	0.0165(5)	0.2918(5)	0	0.0192(16)
O(2)	4*h*, *m*	1.0	0.1804(4)	0.7355(4)	0.5	0.0139(15)
O(3)	8*i*,1	1.0	0.0526(3)	0.4042(3)	0.2791(7)	0.0150(10)
O(4)	8*i*,1	1.0	0.4291(3)	0.5701(4)	0.7326(8)	0.0176(11)
O(5)	8*i*, 1	1.0	0.0392(3)	0.7465(4)	0.3168(7)	0.0184(11)
O(6)	8*i*, 1	1.0	0.1258(3)	0.2425(3)	0.2554(7)	0.0141(10)
O(7)	4*h*, *m*	1.0	0.1292(4)	0.5630(5)	0.5	0.0134(14)
O(8)	8*i*, 1	1.0	0.2653(3)	0.6115(4)	0.6809(7)	0.0189(11)
O(9)	4*h*, *m*	1.0	0.0990(4)	0.8882(5)	0.5	0.0142(15)
O(10)	4*g*, *m*	1.0	0.3262(5)	0.6260(6)	0	0.0187(16)
O(11)w	4*g*, *m*	1.0	0.2446(9)	0.3400(10)	0	0.079(4)

Table 4. Selected interatomic distances for $Rb_{1.66}Cs_{1.34}Tb[Si_{5.43}Ge_{0.57}O_{15}]\cdot H_2O$.

Atoms	Bonds (Å)	Atoms	Bonds (Å)
Tb1O$_6$ Octahedron		**Tb2O$_6$ Octahedron**	
Tb1-O4 x4	2.286(5)	Tb2-O7 ×2	2.262(7)
Tb1-O9 ×2	2.290(7)	Tb2-O3 ×4	2.304(5)
Average	2.287	Average	2.290
(Si,Ge)1O$_4$ Tetrahedron		**(Si,Ge)2O$_4$ Tetrahedron**	
(Si,Ge)1-O9	1.573(7)	(Si,Ge)2-O4	1.582(5)
(Si,Ge)1-O5 ×2	1.632(5)	(Si,Ge)2-O8	1.629(5)
(Si,Ge)1-O2	1.640(7)	(Si,Ge)2-O10	1.637(3)
Average	1.619	(Si,Ge)2-O6	1.662(5)
(Si,Ge)3O$_4$ Tetrahedron		Average	1.628
(Si,Ge)3-O3	1.588(5)	**(Si,Ge)4O$_4$ Tetrahedron**	
(Si,Ge)3-O5	1.617(5)	(Si,Ge)4-O7	1.579(7)
(Si,Ge)3-O1	1.637(3)	(Si,Ge)4-O8 x2	1.634(5)
(Si,Ge)3-O6	1.660(5)	(Si,Ge)4-O2	1.635(7)
Average	1.626	Average	1.621

The new crystal structure of $Rb_{1.66}Cs_{1.34}Tb[Si_{5.43}Ge_{0.57}O_{15}]\cdot H_2O$ consists of mixed (Si, Ge) tetrahedra (Table 3) which are combined into the corrugated layer $[Si_{5.68}Ge_{0.32}O_{15}]_{\infty\infty}$ parallel to *ac* containing four-, six- and eight-membered rings. The isolated TbO_6 centro-symmetric octahedra (Table 4) connected with (Si, Ge) tetrahedral layers into the mixed microporous framework. Rb, Cs atoms and water molecules fill the channels of the framework (Figure 2a,b).

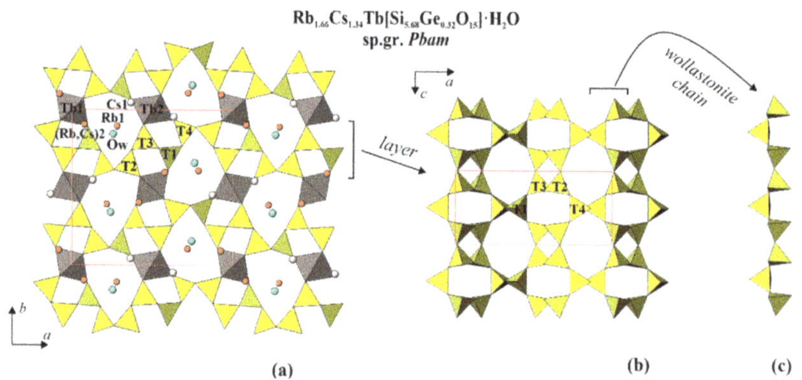

Figure 2. Crystal structure of $Rb_{1.66}Cs_{1.34}Tb[Si_{5.43}Ge_{0.57}O_{15}]\cdot H_2O$: mixed framework in *ab*-projection (**a**), Si-O tetrahedral layer in *ac*-projection (**b**) and isolated wollastonite chain (**c**).

4. Discussion

4.1. Structural Comparison with the Related Layered Silicates

Polytypic relations between isochemical alkali-REE layer silicates and sazhinite were analyzed in [40]. Multiring tetrahedral sheets for four crystal structures: sazhinite $Na_2Ce[Si_6 O_{14}(OH)_2]\cdot nH_2O$ [41], $Na_{2.74}K_{0.26}Ce[Si_6O_{15}]\cdot 2H_2O$ [42], $Na_{2.4}Ce[Si_6O_{15}]\cdot 2H_2O$ [43], $K_3Nd[Si_6O_{15}]\cdot 2H_2O$ [44] were analyzed, emphasizing the presence of a xonotlite-like ribbon. Polytypic relations were derived for the compounds and the crucial role of large K cations was found in forming crystal structures. Some symmetry elements were described: *m* mirror plane for sazhinite structure type, *a* glide for $Na_{2.4}Ce[Si_6O_{15}]\cdot 2H_2O$. Diffraction features, which present some diffusion effects and spurious reflections, were fixed on reciprocal lattice *h0l* and *h1l*.

The crystal structure of a new member of the family, $Rb_{1.66}Cs_{1.34}Tb[Si_{5.43}Ge_{0.57}O_{15}]\cdot H_2O$ is close to the layered $K_3Nd[Si_6O_{15}]\cdot 2H_2O$ [44], $K_3Nd[Si_6O_{15}]$ [45] and $K_3Eu[Si_6O_{15}]\cdot 2H_2O$

(symmetry decrease is caused by the displacement of some atoms from the *m*-plane) [46] (Table 5). All the compounds have identical mixed frameworks consisting of Si-O tetrahedral layers combined with isolated $REEO_6$-octahedra. The new crystal structure differs from these only by the substitution of Rb, Cs for K and the amount of water molecules. Different filling of voids is a characteristic feature of this structural family up to differences in individual crystals of the same mineral sample. Silicate $K_3Eu[Si_6O_{13}(OH)_4]\cdot 2H_2O$ [46] contains ribbon $[Si_6O_{13}(OH)_4]_\infty$ instead of layer (Figure 3a). The similar ribbon as in the latter crystal structure is presented in the layers of all former silicates including the new member. Ribbons are multiplied into the layer along the *a*-axis by inversion center (Figure 3b).

Table 5. The main crystallographic characteristics of the family structures.

Chemical Formula	Space Group	Unit Cell Parameters, Å	Reference
$Rb_{1.66}Cs_{1.34}Tb[Si_{5.43}Ge_{0.57}O_{15}]\cdot H_2O$	*Pbam*	$a = 15.943$ $b = 14.841$ $c = 7.278$	[this work]
$K_3Nd[Si_6O_{15}]\cdot 2H_2O$	*Pbam*	$a = 16.008$ $b = 15.004$ $c = 7.279$	[44]
$K_3Nd[Si_6O_{15}]$	*Pbam*	$a = 16.011$ $b = 14.984$ $c = 7.276$	[45]
$K_3Eu[Si_6O_{15}]\cdot 2H_2O$	$P2_12_12$	$a = 14.852$ $b = 15.902$ $c = 7.243$	[46]
$Na_2Ce[Si_6O_{14}(OH)_2]\cdot nH_2O$ Ce-sazhinite	*Pmm2*	$a = 7.500$ $b = 15.620$ $c = 7.350$	[41]
$Na_3La[Si_6O_{15}]\cdot 2H_2O$ La-sazhinite	*Pmm2*	$a = 7.415$ $b = 15.515$ $c = 7.164$	[47]
β-$K_3Nd[Si_6O_{15}]$	$Bb2_1m$	$a = 14.370$ $b = 15.518$ $c = 14.265$	[44]
$Na_{2.4}Ce[Si_6O_{15}]\cdot 2H_2O$	*Pman*	$a = 7.309$ $b = 14.971$ $c = 7.135$	[43]
$NaNd[Si_6O_{13}(OH)_2]\cdot H_2O$	*Cmm2*	$a = 30.870$ $b = 7.387$ $c = 7.120$	[48]
$NaNd[Si_6O_{15}]\cdot 2H_2O$	*Cmm2*	$a = 7.385$ $b = 30.831$ $c = 7.117$	[49]
$Na_{2.74}K_{0.26}Ce[Si_6O_{15}]\cdot 2H_2O$	*Cmm2*	$a = 7.413$ $b = 30.965$ $c = 7.167$	[42]
$Na_3La[Si_6O_{15}]\cdot 2.25H_2O$	*Cmm2*	$a = 7.415$ $b = 31.008$ $c = 7.153$	[42]
$Na_{2.72}K_{0.25}LaSi_6O_{15}\cdot 2.25H_2O$	*Cmm2*	$a = 7.422$ $b = 31.039$ $c = 7.196$	[42]

The crystal structure of sazhinite $Na_2Ce[Si_6O_{14}(OH)_2]\cdot nH_2O$ [41] ($Na_3La[Si_6O_{15}]\cdot 2H_2O$ [47]) (Table 5) is similar to the Rb,Cs,Tb-silicate-germanate mixed framework and configuration of the corrugated tetrahedral layer with four-, six- and eight-membered rings (Figures 2a and 4a, side projections). However, these layers have a different topology and symmetry visible in frontal projections (Figures 2b and 4b). In sazhinite, the wollastonite chain multiplies into the ribbon by the mirror plane m_y (Figure 4b) and then into the layer by inversion centers valid only for the layer pairs and not for the whole structure. That corresponds to local symmetry operation used in OD theory. A flat ribbon is formed in sazhinite in contrast to a double-decker ribbon in new silicate and in $K_3Eu[Si_6O_{13}(OH)_4]\cdot2H_2O$ because of influence of large K atoms [40]. The β-$K_3Nd[Si_6O_{15}]$ [44] (Table 5) is a distorted variety of the sazhinite crystal structure (Figure 4a–d).

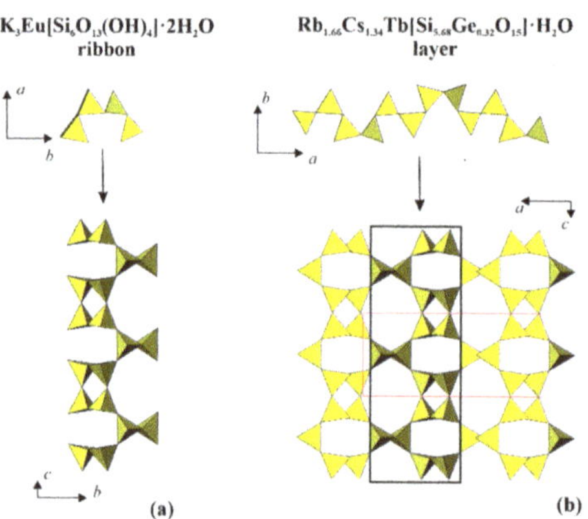

Figure 3. The ribbon in the crystal structure of $K_3Eu[Si_6O_{13}(OH)_4]\cdot2H_2O$ (**a**) and the layer in the crystal structure of $Rb_{1.66}Cs_{1.34}Tb[Si_{5.43}Ge_{0.57}O_{15}]\cdot H_2O$ (**b**).

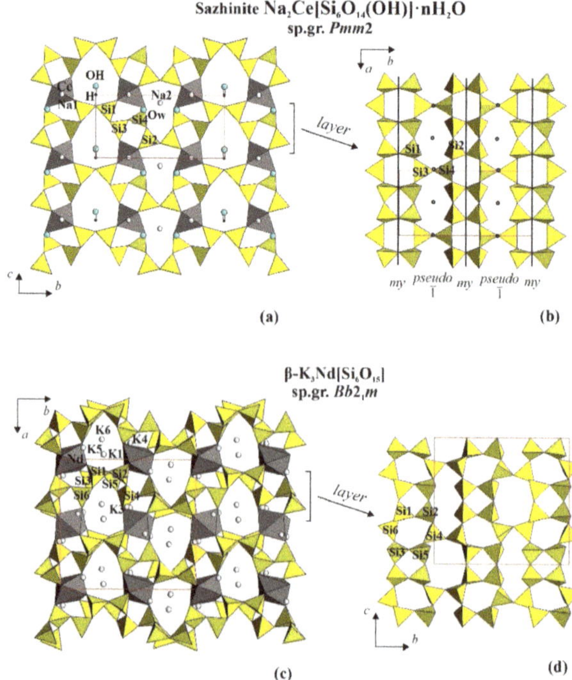

Figure 4. Crystal structure of sazhinite: mixed framework in *bc*-projection (**a**), Si-O tetrahedral layer in *ab*-projection (**b**); crystal structure of β-$K_3Nd[Si_6O_{15}]$: mixed framework in *ab*-projection (**c**), Si-O tetrahedral layer in *bc*-projection (**d**).

The original crystal structure $Na_{2.4}Ce[Si_6O_{15}]\cdot2H_2O$ [43] (Table 5, Figure 5) discussed in [40] has a new variant multiplication of wollastonite chains into the ribbon by a_z glide which are further multiplied by the inversion center in the layer.

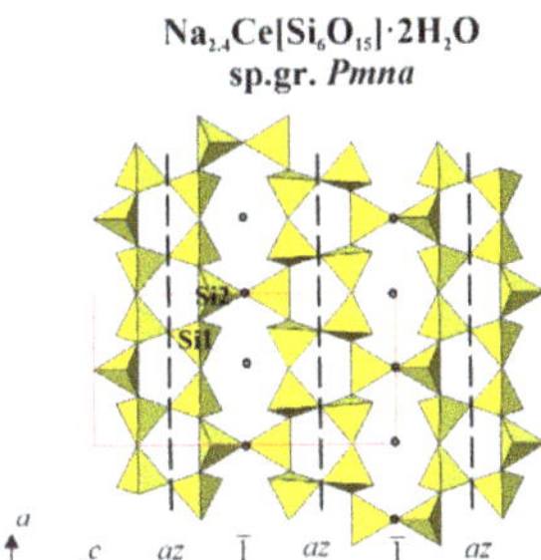

Figure 5. Si-O tetrahedral layer of $Na_{2.4}Ce[Si_6O_{15}]\cdot2H_2O$ crystal structure in *ac*-projection.

In the $NaNd[Si_6O_{13}(OH)_2]\cdot H_2O$ [48] (equal to $Na_3Nd[Si_6O_{15}]\cdot2H_2O$ [49]), $Na_{2.74}K_{0.26}Ce[Si_6O_{15}]\cdot2H_2O$, $Na_3La[Si_6O_{15}]\cdot2.25H_2O$ and $Na_{2.72}K_{0.25}LaSi_6O_{15}\cdot2.25H_2O$ [42] crystal structures (Table 5, Figure 6a,b), two variants of ribbon-forming are observed: by m_x as in sazhinite, and by the b_x operation in accordance with the unit cell selection. Ribbons are connected into the layer by local operation or pseudo-inversion centers $\bar{1}$ (Figure 6b). As a result, the a parameter (b in sazhinite) is doubled (Table 5).

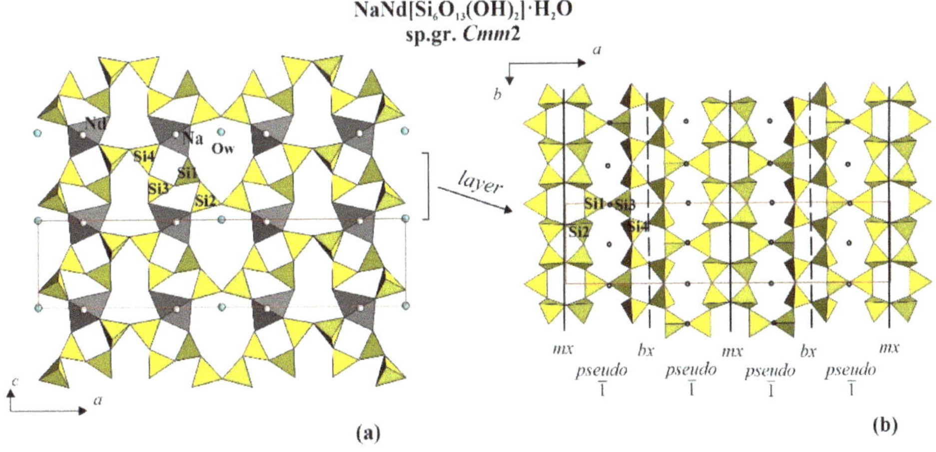

Figure 6. Crystal structure of $NaNd[Si_6O_{13}(OH)_2]\cdot H_2O$: mixed framework in *ac*-projection (**a**), Si-O tetrahedral layer in *ab*-projection (**b**).

4.2. Topology-Symmetry Analysis and Structure Prediction

Based on the topology-symmetry analysis of this family of layered crystal structures and the wollastonite chain with m_x symmetry, we can identify the ribbon from the chain by three variants of symmetry operations: m_y, $\bar{1}$ (equal to 2_x) and a_y. This first step is illustrated at the top of the Figure 7. The ribbons can be connected into the layers in several ways using the same operations: m_y, $\bar{1}$ and a_y left to right, shown on the next line marked as "layers". The first case is when the ribbon with the symmetry Pm_y is multiplied by the mirror plane m_y giving the layer with the polar symmetry group $Pmm2$. Multiplication of the same ribbon by the inversion center gives the layer of idealized sazhinite with symmetry group $Pmmb$. The second case is when a centro-symmetric ribbon with symmetry $P\bar{1}$ is multiplied by m_y, giving the *Pmmb* space group or by the inversion center giving the *P2/m* space group The third case describes multiplying of the ribbon with another symmetry group *Pa* by m_y or inversion center giving the *Cmm2* or *Pman* space groups, correspondingly. The latter variant presents known structure $Na_{2.4}Ce[Si_6O_{15}]\cdot2H_2O$ (space group setting *Pmna*) [43] In all drawings, the resulting unit cells are shown.

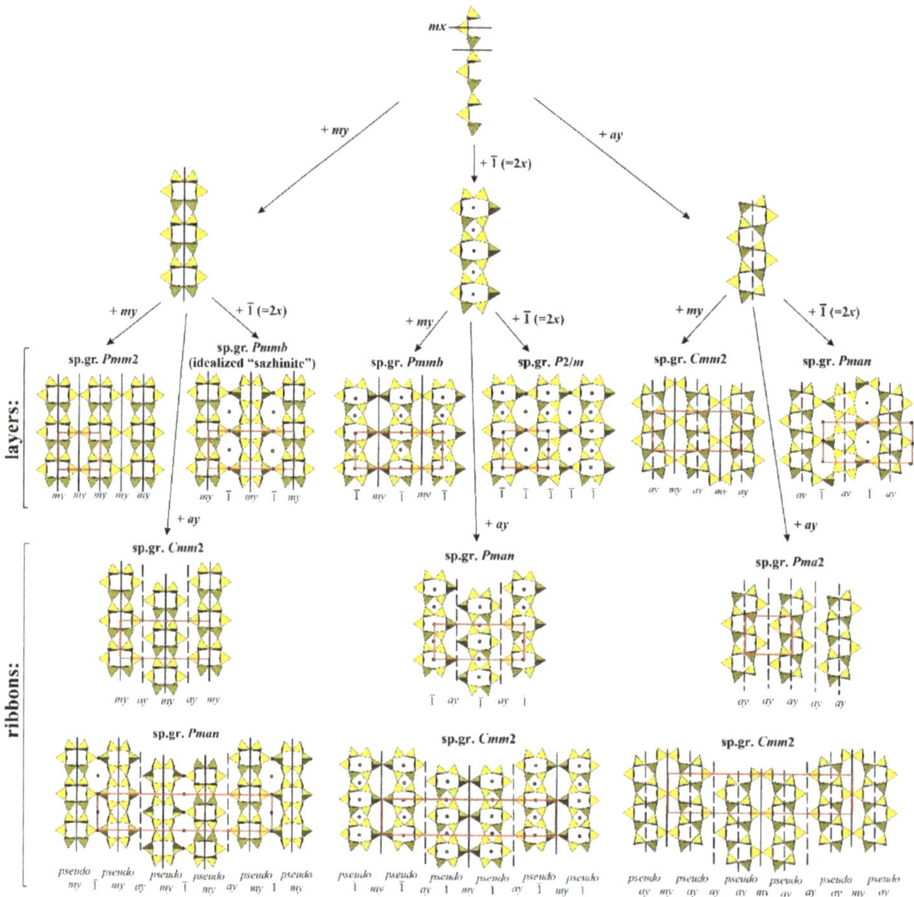

Figure 7. Construction of anionic Si-O radicals based on wollastonite chain and different multiplying symmetry operations.

Structural diversity has an OD character described by a symmetry groupoid. The chain, which has a one-dimensional periodicity, is multiplied in the ribbon in different ways; thus, a groupoid family symbol of lower rank is used as it was in [25] for borates:

$$P\,(m_x)\,1 \qquad \lambda\text{-PO,}$$

$$\overline{1}||m_y||a_y \qquad \sigma\text{-PO.}$$

Different ribbons are joined by different symmetrical operations in different layers, and that corresponds to the groupoid family symbol:

$$(Pmm2||P2/m||Pma2)\,1 \qquad \lambda\text{-PO,}$$

$$\overline{1}||m_y||a_y \qquad \sigma\text{-PO.}$$

All variants are shown in Figure 7 and belong to the members of the unify OD family with the maximum degree of order (MDO) because the λ-PO and σ-PO operations are the same for every layer, and all the ribbon pairs in the layer are equal.

Periodic crystal structures with the alternation of λ-PO and σ-PO may exist in the family up to disordered members if no order will be in initial λ-PO or σ-PO.

It is possible to predict not only layers but also different ribbons containing crystal structures, if we use a_y as σ-PO. They are shown in the next row in Figure 7 for different initial ribbon λ-PO (see arrows) with the different σ-PO multiplication and resulting space groups and unit cells. Hypothetic crystal structures with double ribbons are shown at the bottom of Figure 7 as an intermediate case between ribbon and layered crystal structures.

The specific diffraction effects described in [40] were explained by the polytypic nature of the compounds. This is consistent with the order–disorder (OD) nature of the crystal structures belonging to the OD family described using topology-symmetry analysis.

5. Conclusions

Crystals of $Rb_{1.66}Cs_{1.34}Tb[Si_{5.43}Ge_{0.57}O_{15}]\cdot H_2O$ have been synthesized hydrothermally in a multi-component system at T = 280 °C and P = 100 atm and luminescence properties on the crystals were investigated. The crystal structure of new Rb,Cs,Tb-silicate-germanate with the isomorphic substitution in tetrahedra consists of corrugated layers $[Si_{5.43}Ge_{0.57}O_{15}]_{\infty\infty}$ which are connected with isolated TbO_6-octahedra into the mixed microporous framework. Larger than Na,K-atoms, Rb, Cs atoms and water molecules fill the channels of the framework. The substitution of larger Nd, Eu REE by smaller Tb REE does not changes the dimensions of voids. Structural comparison with the related layered silicates was carried out. Modular description added with the symmetry analysis of the OD approach allows to systematically describe the structural families or to predict new members. Topology-symmetry analysis of this silicates family was performed based on consideration of the wollastonite chain and symmetry variants of chain combination into the ribbons and then into the layers. The structural representatives of minerals and synthetic compounds are potentially extended. One example of "hypothetical structure" with the space group *Pman* corresponds to real $Na_{2.4}Ce[Si_6O_{15}]\cdot 2H_2O$ being in a systematically derived position in the field of structural variants. Most of the hypothetical layered and ribbon crystal structures are not yet discovered. Their description will help to recognize new crystal structures of minerals in nature or in synthetic experiments and to confirm the predicted models.

Author Contributions: Conceptualization, A.T., E.B. and O.D.; methodology, A.T., E.B., O.D., A.V. and D.D.; validation, E.B. and O.D.; investigation, A.T., A.V. and D.D.; writing—original draft preparation, A.T., E.B., O.D., A.V. and D.D.; writing—review and editing, A.T. and E.B.; visualization, A.T.; supervision, E.B. All authors have read and agreed to the published version of the manuscript.

Funding: Investigation of luminescence properties was supported by RF President Scholarship (SP-859.2021.1).

Data Availability Statement: The data supporting reported results can be found as CCDC CSD 2062486 contains crystallographic data for this paper, www.ccdc.cam.ac.uk/data_request/cif.

Acknowledgments: The authors are grateful to Vasiliy Yapaskurt for determination of the chemical composition, to Natalia Zubkova for diffraction experimental data and to the Editor (G. Ferraris) for valuable remarks and suggestions.

Conflicts of Interest: The authors declare no conflict of interest.

References

1. Mineralogy Database. Available online: http://www.mindat.org/ (accessed on 20 October 2020).
2. ICSD FIZ. Available online: http://www.fiz-karlsruhe.de (accessed on 20 October 2020).
3. Crystallography Open Database. Available online: http://www.crystallography.net (accessed on 20 October 2020).
4. Garra, W.; Marchetti, F.; Merlino, S. Tb/Na tobermorite: Thermal behaviour and high temperature products. *J. Solid State Chem.* **2009**, *182*, 1529–1532. [CrossRef]

5. Wierzbicka-Wieczorek, M.; Göckeritz, M.; Kolitsch, U.; Lenz, C.; Giester, G. Crystallographic and Spectroscopic Investigations o Nine Metal-Rare-Earth Silicates with the Apatite Structure Type. *Eur. J. Inorg. Chem.* **2015**, *2015*, 948–963. [CrossRef]

6. Latshaw, A.M.; Wilkins, B.O.; Hughey, K.D.; Yeon, J.; Williams, D.E.; Tran, T.T.; Halasyamani, P.S.; Loye, H.-C.Z. A 5 RE 4 X [T(4] 4 crystal growth and photoluminescence. Fluoride flux synthesis of sodium and potassium rare earth silicate oxyfluoride *CrystEngComm* **2015**, *17*, 4654–4661. [CrossRef]

7. Fleet, M.E.; Liu, X. Rare earth disilicates R2Si2O7 (R = Gd, Tb, Dy, Ho): Type B. *Z. Krist. Cryst. Mater.* **2003**, *218*, 795–80 [CrossRef]

8. Vidican, I.; Smith, M.D.; Zur Loye, H.-C. Crystal growth, structure determination and optical properties of new potassi-um-rare earth silicates K3RESi2O7 (RE = Gd, Tb, Dy, Ho, Er, Tm, Yb, Lu). *J. Solid St. Chem.* **2003**, *170*, 203–210. [CrossRef]

9. Sieke, C.; Hartenbach, I.; Schleid, T. Sulfidisch derivatisierte Oxodisilicate der schweren Lanthanide vom Formeltyp $M_4S_3(Si_2O_7$ (M = Gd − Tm). *Z. Nat. B J. Chem. Sci.* **2002**, *B57*, 1427–1432.

10. Ananias, D.; Kostova, M.; Paz, F.A.A.; Ferreira, A.; Carlos, L.D.; Klinowski, J.; Rocha, J. Photoluminescent Layered Lanthanid Silicates. *J. Am. Chem. Soc.* **2004**, *126*, 10410–10417. [CrossRef]

11. Fulle, K.; Sanjeewa, L.D.; McMillen, C.D.; Kolis, J.W. Crystal chemistry and the role of ionic radius in rare earth tetrasilicate $Ba_2RE_2Si_4O_{12}F_2(RE = Er^{3+}–Lu^{3+})$ and $Ba_2RE_2Si_4O_{13}(RE = La^{3+}–Ho^{3+})$. *Acta Crystallogr. Sect. B Struct. Sci. Cryst. Eng. Mate* **2017**, *73*, 907–915. [CrossRef]

12. Topnikova, A.P.; Belokoneva, E.L.; Dimitrova, O.V.; Volkov, A.S.; Nelyubina, Yu.V. $Na_3Tb_3[Si_6O_{18}]\cdot H_2O$, a synthetic analogue o microporous mineral gerenite. *Cryst. Rep.* **2016**, *61*, 566–570. [CrossRef]

13. Lee, C.-S.; Liao, Y.-C.; Hsu, J.-T.; Wang, S.-L.; Lii, K.-H. $Rb_2REGaSi_4O_{12}$ (RE = Y, Eu, Gd, Tb): Luminescent Mixed-Anion Doubl Layer Silicates Containing Chains of Edge-Sharing REO_7 Pentagonal Bipyramids. *Inorg. Chem.* **2008**, *47*, 1910–1912. [CrossRef]

14. Bao, X.; Liu, X.; Liu, X. High-pressure synthesis, crystal structure and photoluminescence properties of a new terbium silicate $Na_2Tb_{1.08}Ca_{2.92}Si_6O_{18}H_{0.8}$. *RSC Adv.* **2017**, *7*, 50195. [CrossRef]

15. Wang, G.; Li, J.; Yu, J.; Chen, P.; Pan, Q.; Song, H.; Xu, R. $Na_3TbSi_3O_9\cdot 3H_2O$: A New Luminescent Microporous Terbium(III Silicate Containing HelicalSechserSilicate Chains and 9-Ring Channels. *Chem. Mater.* **2006**, *18*, 5637–5639. [CrossRef]

16. Morrison, G.; Latshaw, A.M.; Spagnuolo, N.R.; Loye, H.-C.Z. Observation of Intense X-ray Scintillation in a Family of Mixec Anion Silicates, $Cs_3RESi_4O_{10}F_2$(RE = Y, Eu–Lu), Obtained via an Enhanced Flux Crystal Growth Technique. *J. Am. Chem. Soc* **2017**, *139*, 14743–14748. [CrossRef]

17. Taroev, V.K.; Kashaev, A.A.; Malcherek, T.; Goettlicher, J.; Kaneva, E.V.; Vasiljev, A.D.; Suvorova, L.F.; Suvorova, D.S.; Tauson, V.L Crystal structures of new potassium silicates and aluminosilicates of Sm, Tb, Gd, and Yb and their relation to the armstrongite $(CaZr(Si_6O_{15})\cdot 3H_2O)$ structure. *J. Solid State Chem.* **2015**, *227*, 196–203. [CrossRef]

18. Zhao, X.; Li, J.; Chen, P.; Li, Y.; Chu, Q.; Liu, X.; Yu, J.; Xu, R. New Lanthanide Silicates Based on Anionic Silicate Chain, Layer and Framework Prepared under High-Temperature and High-Pressure Conditions. *Inorg. Chem.* **2010**, *49*, 9833–9838. [CrossRef] [PubMed]

19. Ananias, D.; Ferreira, A.; Rocha, J.; Ferreira, P.; Rainho, J.P.; Morais, C.; Carlos, L.D. Novel Microporous Europium and Ter-bium Silicates. *J. Am. Chem. Soc.* **2001**, *123*, 5735–5742. [CrossRef] [PubMed]

20. Liebau, F. *Structural Chemistry of Silicates: Structure, Bonding, and Classification*; Springer: Berlin/Heidelberg, Germany, 1985; 347p

21. Pushcharovsky, D.Y. *Structural Mineralogy of Silicates and Their Synthetic Analogues*; Nedra: Moscow, Russia, 1986; 160p.

22. Ferraris, G.; Makovicky, E.; Merlino, S. *Crystallography of Modular Materials*; Oxford University Press (OUP): Oxford, UK, 2008.

23. Dornberger-Schiff, K. *Grundzuege Einer Theorie der OD-Strukturen aus Schichten*; Deutsche Akademie der Wissenschaften, Berlin Abhandlungen, Klasse fur Chemie, Geologie and Biologie: Halle, Germany, 1964; Volume 3, pp. 1–106.

24. Merlino, S. OD Structures in Mineralogy. *Per. Mineral.* **1990**, *59*, 69–92.

25. Belokoneva, E.L. Borate crystal chemistry in terms of the exnetded OD theory: Topology and symmetry analysis. *Cryst. Rev.* **2005** *11*, 151–198. [CrossRef]

26. Ivanova, A.G.; Belokoneva, E.L.; Dimitrova, O.V. New condensed acid diborate $GdH[B_2O_5]$ with chain radical $[B2\square B2\Delta O10]8-]\infty$ Synthesis and crystal structure; diborates and their structural system in terms of OD theory. *Russ. J. Inorg. Chem.* **2004**, *49* 816–822.

27. Belokoneva, E.L.; Reutova, O.V.; Dimitrova, O.V.; Volkov, A.S. Germanosilicate $Cs_2In_2[(Si_{2.1}Ge_{0.9})_2O_{15}](OH)_2\cdot H_2O$ with a New Corrugated Tetrahedral Layer: Topological Symmetry-Based Prediction of Anionic Radicals. *Crystallogr. Rep.* **2020**, *65*, 566–572 [CrossRef]

28. Dorenbos, P. Exchange and crystal field effects on the 4fn 15d levels of Tb3. *J. Phys. Condens. Matter* **2003**, *15*, 6249–6268. [CrossRef]

29. Pisarski, W.A.; Zur, L.; Sołtys, M.; Pisarska, J. Terbium-terbium interactions in lead phosphate glasses. *J. Appl. Phys.* **2013**, *113* 143504. [CrossRef]

30. Berdowski, P.A.M.; Lammers, M.J.J.; Blasse, G. 5D3-5D4 cross-relaxation of Tb^{3+} in α-GdOF. *Chem. Phys. Lett.* **1985**, *113*, 387–390

31. Van Uitert, L.G.; Johnson, L.F. Energy Transfer between Rare—Earth Ions. *J. Chem. Phys.* **1966**, *44*, 3514. [CrossRef]

32. Tonooka, K.; Nishimura, O. Spectral changes of Tb^{3+} fluorescence in borosilicate glasses. *J. Lumin.* **2000**, *87*, 679–681. [CrossRef]

33. Deyneko, D.V.; Morozov, V.A.; Vasin, A.A.; Aksenov, S.M.; Dikhtyar, Y.Y.; Stefanovich, S.Y.; Lazoryak, B.I. The crystal site engineering and turning of cross-relaxation in green-emitting β-$Ca_3(PO_4)_2$-related phosphors. *J. Lumin.* **2020**, *223*, 117196 [CrossRef]

34. *CrysAlis PRO*; Agilent Technologies Ltd.: Yarnton, Oxfordshire, UK, 2014.

35. Sheldrick, G.M. A short history of SHELX. *Acta Cryst.* **2008**, *64*, 112–122. [CrossRef] [PubMed]
36. Farrugia, L.J. WinGX and ORTEP for Windows: An update. *J. Appl. Cryst.* **2012**, *45*, 849. [CrossRef]
37. Pauling, L. *The Nature of the Chemical Bond*; Cornell University: Ithaca, NY, USA, 1960; 644p.
38. Sheldrik, G.M. Crystal structure refinement with SHELXL. *Acta Cryst.* **2015**, *71*, 3–8.
39. Dowty, E. *ATOMS*; Shape Software: Kingsport, TN, USA, 2006.
40. Cadoni, M.; Ferraris, J. Polytypic and polymorphic relations between sazhinite and isochemical alkali-REE layer silicates. *Eur. J. Mineral.* **2011**, *23*, 85–90. [CrossRef]
41. Shumyatskaya, N.G.; Voronkov, A.A.; Pyatenko, Yu.A. Sazhinite $Na_2Ce[Si_6O_{14}(OH)] \cdot nH_2O$, a new member of crystal chemical family of dalyite. *Sov. Phys. Cryst.* **1980**, *25*, 728–734.
42. Cadoni, M.; Cheah, Y.L.; Ferraris, G. New RE microporous heteropolyhedral silicates containing 41516182 tetrahedral sheets. *Acta Crystallogr. Sect. B Struct. Sci.* **2010**, *66*, 158–164. [CrossRef] [PubMed]
43. Jeong, H.-K.; Chandrasekaran, A.; Tsapatsis, M. Synthesis of a new open framework cerium silicate and its structure determination by single crystal X-ray diffraction. *Chem. Commun.* **2002**, 2398–2399. [CrossRef] [PubMed]
44. Haile, S.M.; Wuensch, B.J. Structure, phase transitions and ionic conductivity of $K_3NdSi_6O_{15} \cdot xH_2O$. II. Structure of β-$K_3NdSi_6O_{15}$. *Acta Cryst.* **2000**, *56*, 349–362. [CrossRef]
45. Pushcharovsky, D.Y.; Karpov, O.G.; Pobedimskaya, E.A.; Belov, N.V. Crystal structure of $K_3NdSi_6O_{15}$. *Dokl. AN SSSR* **1977**, *234*, 1323–1326.
46. Rastsvetaeva, R.K.; Aksenov, S.M.; Taroev, V.K. Crystal Structures of Endotaxic Phases in Europium Potassium Silicate Having a Pellyite Unit Cell. *Crystallogr. Rep.* **2010**, *55*, 1041–1049. [CrossRef]
47. Cámara, F.; Ottolini, L.; Devouard, B.; Garvie, L.A.J.; Hawthorne, F.C. Sazhinite-(La), $Na_3LaSi_6O_{15}(H_2O)_2$, a new mineral from the Aris phonolite, Namibia: Description and crystal structure. *Miner. Mag.* **2006**, *70*, 405–418. [CrossRef]
48. Karpov, O.G.; Pushcharovsky, D.Y.; Pobedimskaya, E.A.; Burshtein, I.F.; Belov, N.V. Crystal structure of rare earth silicate $NaNdSi_6O_{13}(OH)_2 \cdot nH_2O$. *Dokl. AN SSSR* **1977**, *236*, 593–596.
49. Haile, S.M.; Wuensch, B.J.; Laudise, R.A.; Maier, J. Structure of $Na_3NdSi_6O_{15} \cdot 2H_2O$—A Layered Silicate with Paths for Possible Fast-Ion Conduction. *Acta Cryst.* **1997**, *53*, 7–17. [CrossRef]

Article

A Mero-Plesiotype Series of Vanadates, Arsenates, and Phosphates with Blocks Based on Densely Packed Octahedral Layers as Repeating Modules

Olga Yakubovich [1,*] and Galina Kiriukhina [1,2]

1 Department of Crystallography, Geological Faculty, Lomonosov Moscow State University, 119991 Moscow, Russia; g-biralo@yandex.ru
2 Institute of Experimental Mineralogy RAS, 142432 Chernogolovka, Russia
* Correspondence: yakubol@geol.msu.ru; Tel.: +7-903-975-91-06

Abstract: The family of layered vanadates, arsenates, and phosphates is discussed in terms of a modular concept. The group includes minerals vésignéite and bayldonite, and a number of synthetic analogous and modifications which are not isotypic, but their crystal structures comprise similar blocks (modules) consisting of a central octahedral layer filled by atoms of d elements (Mn, Ni, Cu, or Co) and adjacent $[VO_4]$, $[AsO_4]$, or $[PO_4]$ tetrahedra. The octahedral layers are based on the close-packing of oxygen atoms. Within these layers having the same anionic substructure, the number and distribution of octahedral voids are different. In the crystal structures of compounds participating in the polysomatic series, these blocks alternate with various other structural fragments. These circumstances define the row of structurally-related vanadates, arsenates, and phosphates as a mero-plesiotype series. Most of the series members exhibit magnetic properties, representing two-dimensional antiferromagnets or frustrated magnets.

Keywords: modular structures; polysomes (series); synthetic analogues of minerals; transition metal phosphates; X-ray diffraction; antiferromagnets; frustrated magnets; kagomé lattice

check for **updates**

Citation: Yakubovich, O.; Kiriukhina, G. A Mero-Plesiotype Series of Vanadates, Arsenates, and Phosphates with Blocks Based on Densely Packed Octahedral Layers as Repeating Modules. *Minerals* **2021**, *11*, 273. https://doi.org/10.3390/min11030273

Academic Editor: Giovanni Ferraris

Received: 29 January 2021
Accepted: 4 March 2021
Published: 7 March 2021

Publisher's Note: MDPI stays neutral with regard to jurisdictional claims in published maps and institutional affiliations.

1. Introduction

The modular approach for interpreting the crystal structures of minerals is well known. Its intensive use in modern crystal chemistry has become possible after the pioneer works of Thomson [1–3] and Veblen [4]. The modular concept operates with large segments (modules), which represent stable polyhedral complexes of definite topology and size, and can differ in structure and/or composition. Within this concept, the crystal structure can be interpreted as a derivative of diverse modules; then it is meant as a polysome. The polysomatism considers a particular crystal structure in required connection with other structures, assembled from the same modules. This method allows distinguishing series of non-isotypic, but crystal chemically-related phases, so-called polysomatic series, with structures built by fragments of the same topology taken in various combinations. The stability of repeating fragments is controlled by the energetic advantage of the associations of polyhedra forming these modules. Obviously, one can use the modular approach only in the case of several compounds (at least two) containing similar substructural segments (modules), with clearly pronounced rules for their spatial arrangement and alternation.

To expand the field of structurally-related compounds within the polysomatic series, Makovicky [5] proposed to adopt a possibility of variations both in the structure and in the chemical composition of modules still keeping their topological similarity. These advanced sequences of compounds were called mero-plesiotype series. In the framework of this concept, a "common" fragment for all of the structures shows clear variability; besides, modules of another type are dissimilar for different members of the series [6,7].

Through efforts of numerous scientific schools, this methodology has allowed the introduction of many families of inorganic compounds, including minerals, as polysomatic series of oxides, sulfosalts, silicates, manganates, phosphates, etc. [8–21]. Several aspects of modular analysis such as symmetrical, topological classification can be mentioned. Usually they are all interconnected; all are presented and complement each other in the study. In addition, the polysomatic model facilitates the representation of mineral transformation and substitution reactions and also allows predicting possible topological constraints in the propagation of mineral reactions [22]. It is difficult to overestimate the heuristic potential of this approach, since the prediction of the structural state of atoms and atomic groups, the way of their interaction within the crystal, ion migration paths, and the features of chemical bonds are extremely important in the context of the targeted selection of compounds for studying their physical properties that underlie the creation of new materials.

Within a program to explore the synthesis and crystal chemistry of compounds, potentially interesting as possible battery electrodes and/or magnetic materials, we have established crystal structures of two transition and alkaline metal phases synthesized under hydrothermal conditions, namely the first vanadate carbonate, $K_2Mn_3[VO_4]_2(CO_3)$ [23] and the sodium nickel hydroxide phosphate, $Na_2Ni_3(OH)_2(PO_4)_2$ [24]. A structural study of the $K_2Mn_3[VO_4]_2(CO_3)$ was carried out on a crystal showing [110] twinning by mero-hedry. [110] is a twofold axis of the lattice (point group $6/mmm$) but not of the structure (point group $6/m$) and can act as twinning operation. Both structures are built of similar slabs consisting of a central octahedral layer partially filled by atoms of d elements Mn/Ni and adjacent [VO_4] or [PO_4] tetrahedra. A noticed similarity of their crystal architecture completed with research on structurally-related synthetic phases and minerals allowed us to establish a new polysomatic series of vanadates, arsenates, and phosphates with densely packed octahedral layers decorated by tetrahedra, as repeating fragments.

2. The Main Module Topology

The core structural module upon which the whole family is built presents the block centered by an octahedral layer filled by atoms of d elements Mn/Ni/Cu/Co and adjacent from both sides [VO_4], [PO_4], and [AsO_4] tetrahedra. (In all the following figures, Mn, Ni and Cu atoms are shown in pink, green, and turquoise colors, respectively; P-, As- and V-centered tetrahedra are colored yellow, blue, and olive green). The central layer is based on a closest sphere packing of oxygen atoms. The octahedral voids within the layer may be differently populated by the cations; therefore, various arrangements of filled and empty octahedra arise. Sheets of the brucite type (the so-called trioctahedral layers) are the densest, since they have fully occupied MeO_6 octahedra sharing edges. If $^2/_3$ of the voids are filled, a gibbsite sheet (dioctahedral) is formed. These two varieties of layers present essential structural fragments of phyllosilicates and clay minerals. In the crystal structures under consideration, a fraction of the empty octahedral voids inside the oxygen sphere packing can be equal to $^1/_3$ or $^1/_4$. Besides, these voids can be differently distributed within the layer, causing differences in the structure design. From both sides of the empty octahedral voids, the tetrahedral groups are attached by sharing three vertices with O atoms of the layer, while the apical vertex is pointing outside towards the next module that is individual for each structure, thus defining the polysomatic series as the mero-plesiotype row. The formulae of the basic blocks may be written as $\{Me_2[TO_4]_2\}$, or $\{Me_3(OH)_2[TO_4]_2\}$, according to the number of the empty octahedra in the layer.

It is worth mentioning that the crystal structures of minerals reppiaite, $Mn_5(OH)_4[VO_4]_2$ [25], cornubite, $Cu_5(OH)_4[AsO_4]_2$ [26], and their synthetic arsenate formula analogue $Ni_5(OH)_4[AsO_4]_2$ [27] are also based on the closest sphere packing of O atoms. In all three cases, the atoms of transition metals occupy $^5/_6$ of the octahedral voids (Figures 1 and 2). However, the way of the empty octahedra arrangement in the monoclinic reppiaite and isotypic Ni arsenate structures on one hand, and in the structure of triclinic cornubite on the other, are different, providing a diverse topology of their cationic substructures, as dual-width stripes of the triangular net separated by honeycombs or dual-width stepped stripes of the triangular net separated by honeycombs (Figure 3). Note that both arrangements, as well as other diagrams shown in Figure 3 may be obtained as vacancy-modified triangular nets. The tetrahedral groups are placed on both sides of the layer just above and under empty octahedra with the formation of the $\{Me_5(OH)_4[TO_4]_2\}$ block. In all three structures, the similar blocks are stacked together in a direction normal to the plane of the layer through oxygen-bridging contacts. The neighboring modules are semi-translationally displaced in one direction to form an AA′ sequence with a period value of about 9.4 Å (Table 1). Accordingly, it can be concluded that these phases with the crystal structures built exclusively from basic modules, on which the polysomatic series of vanadates, arsenates, and phosphates is based, represent the archetype structures.

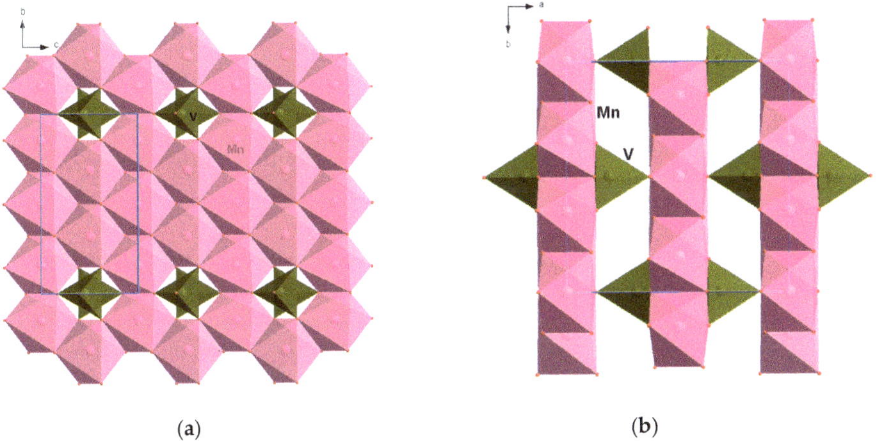

(a) (b)

Figure 1. The crystal structure of reppiaite, $Mn_5(OH)_4[VO_4]_2$ in yz (**a**) and xy (**b**) projections.

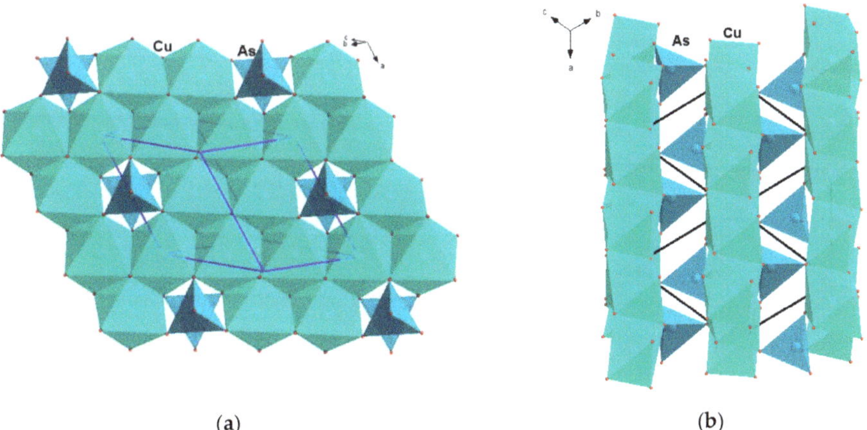

(a) (b)

Figure 2. The crystal structure of cornubite, $Cu_5(OH)_4[AsO_4]_2$ in projections along the [011] (**a**) and [111] (**b**) directions.

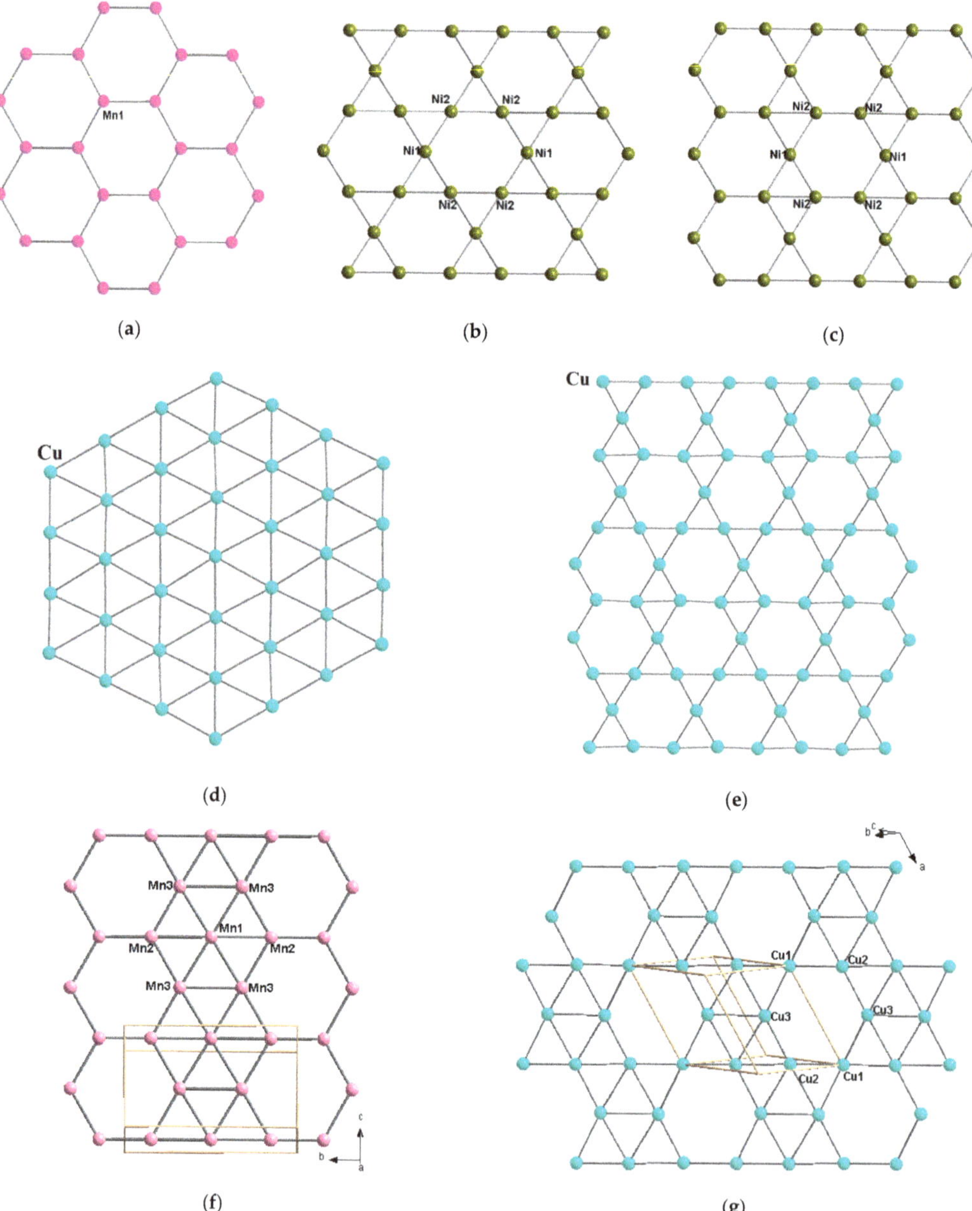

Figure 3. The topology of magnetic subsystems (cationic substructures): the honeycomb-type (or graphite-type) lattice characteristic of the gibbsite layer (the cationic substructure of the core modules in vanadate carbonates, $BaNi_2[VO_4]_2$ and other double vanadates, arsenates and phosphates) (**a**); The kagomé lattice (bayldonite, $Pb(Cu,Zn)_3(OH)_2[AsO_4]_2$ and vésignéite, $BaCu_3(OH)_2[VO_4]_2$) (**b**); The stripes of the triangular net separated by honeycombs ($Na_2Ni_3(OH)_2[PO_4]_2$ and $K_2Mn_3(OH)_2[VO_4]_2$) (**c**); The triangular lattice inherent for the brucite-type layers (the cationic substructure of the B module in $Cu_{13}(OH)_{10}F_4[VO_4]_4$) (**d**); The stepped stripes of the triangular net separated by honeycombs ($Cu_{13}(OH)_{10}F_4[VO_4]_4$) (**e**); The dual-width stripes of the triangular net separated by honeycombs (reppiaite, $Mn_5(OH)_4[VO_4]_2$) (**f**); The dual-width stepped stripes of the triangular net separated by honeycombs (cornubite, $Cu_5(OH)_4[AsO_4]_2$) (**g**).

Table 1. Mero-plesiotype series of vanadates, arsenates, and phosphates of the first-row transition metals.

Mineral/ Synthetic Phase Ref.	Unit-cell Parameters a, b, c (Å) and Angles α, β, γ (°)	Space Group, V (Å^3), Z	Fraction of Filled Octahedra, Layer Topology, Module Sequence	Magnetic Behavior
		Phases with archetype crystal structure		
Reppiaite Mn$_5$(OH)$_4$[VO$_4$]$_2$ [25]	a **9.604(2)** b 9.558(2) β 98.45(1) c 5.393(1)	$C2/m$ 489.7 2	$^5/_6$, dual-width triangular stripes separated by honeycombs, (AA′)	Canted antiferromagnetic ordering below 57 K
Ni$_5$(OH)$_4$[AsO$_4$]$_2$ [27]	a **9.291(2)** b 9.008(2) β 98.70(3) c 5.149(1)	$C2/m$ 426.0 2	$^5/_6$, dual-width triangular stripes separated by honeycombs, (AA′)	——
Cornubite Cu$_5$(OH)$_4$[AsO$_4$]$_2$ [26]	a 6.121(1) α 92.93(1) b 6.251(1) β 111.30(1) c 6.790(1) γ 107.47(1)	$P\bar{1}$ 227.1 1	$^5/_6$, dual-width triangular stepped stripes separated by honeycombs, (AA′)	——
		Compounds forming the mero-plesiotype series		
K$_2$Mn$_3$[VO$_4$]$_2$(CO$_3$) [23]	a 5.201(1) c **22.406(3)**	$P6_3/m$ 524.9 2	$^2/_3$, honeycomb (ABA′B′)	The honeycomb substructure orders antiferromagnetically at 85 K; the triangular substructure displays two ordered states at 3 and 2.2 K
K$_2$Co$_3$[VO$_4$]$_2$(CO$_3$) [28]	a 5.0931(2) c **22.1551(13)**	$P6_3/m$ 497.7 2	$^2/_3$, honeycomb (ABA′B′)	Canted antiferromagnetic ordering below 8 K
Rb$_2$Mn$_3$[VO$_4$]$_2$(CO$_3$) [28]	a 5.2488(3) c **22.7020(14)**	$P\bar{3}$ 1c 541.6 2	$^2/_3$, honeycomb (ABA′B′)	The honeycomb substructure orders antiferromagnetically at 77 K; the triangular substructure exhibits two transitions at 2.3 K and 1.5 K
BaNi$_2$[VO$_4$]$_2$ [29]	a 5.028(1) c **22.345(3)**	$R\bar{3}$ 489.4 3	$^2/_3$, honeycomb (ABA′B′A″B″)	Antiferromagnetic long-range ordering close to 50 K
BaCo$_2$[PO$_4$]$_2$ [30]	a 4.8554(6) c **23.2156(17)**	$R\bar{3}$ 474.0 3	$^2/_3$, honeycomb (ABA′B′A″B″)	Competing short range magnetic orders below T_{N1} ~ 6 K and T_{N2} ~ 3.5 K.
BaCo$_2$[AsO$_4$]$_2$ * [31]	a 5.007(1) c **23.491(5)**	$R\bar{3}$ 510.0 3	$^2/_3$, honeycomb (ABA′B′A″B″)	Frustrated magnet

Table 1. *Cont.*

Mineral/ Synthetic Phase Ref.	Unit-cell Parameters a, b, c (Å) and Angles α, β, γ (°)	Space Group, V (Å^3), Z	Fraction of Filled Octahedra, Layer Topology, Module Sequence	Magnetic Behavior
Compounds forming the mero-plesiotype series				
NaNi[AsO$_4$] [32]	a 4.955(3) c **26.47(3)**	$R\bar{3}$ 562.8 6	$^2/_3$, honeycomb (ABA'B'A''B'')	—
KNi[AsO$_4$] [32]	a 4.97208(2) c **28.52606(10)**	$R\bar{3}$ 610.7 6	$^2/_3$, honeycomb (ABA'B'A''B'')	—
Vésignéite BaCu$_3$(OH)$_2$[VO$_4$]$_2$ [33]	a 10.270(2) b 5.911(1) β 116.42(3) c **7.711(2)**	$C2/m$ 419.2 2	$^3/_4$, kagomé (AB)	Strong antiferromagnetic interactions, no long-range order down to 16 K
BaNi$_3$(OH)$_2$[VO$_4$]$_2$ [34]	a 10.213(6) b 5.816(3) β 117.01(4) c **7.888(4)**	$C2/m$ 417.4 2	$^3/_4$, kagomé (AB)	Glassy transition at 19 K, magnetic frustration from a competition between ferro- and antiferromagnetic ordering
Na$_2$Ni$_3$(OH)$_2$[PO$_4$]$_2$ [24]	a **14.259(5)** b 5.695(2) β 104.28(3) c 4.933(1)	$C2/m$ 388.2 2	$^3/_4$, triangular stripes separated by honeycombs, (ABA'B')	Antiferromagnetic ordering at 33.4 K
Na$_2$Co$_3$(OH)$_2$[VO$_4$]$_2$ [35]	a **14.5847(11)** b 5.9552(4) β 104.068(2) c 5.1414(4)	$C2/m$ 433.2 2	$^3/_4$, triangular stripes separated by honeycombs, (ABA'B')	Antiferromagnetic ordering at 4.4 K
K$_2$Mn$_3$(OH)$_2$[VO$_4$]$_2$ [36]	a **15.204(2)** b 6.159(1) β 105.40(1) c 5.400(1)	$C2/m$ 487.5 2	$^3/_4$, triangular stripes separated by honeycombs, (ABA'B')	Antiferromagnetic ordering at 50 K
Bayldonite Pb(Cu,Zn)$_3$(OH)$_2$ [AsO$_4$]$_2$ [37]	a 10.147(2) b 5.892(1) β 106.05(1) c **14.081(2)**	$C2/c$ 809.0 4	$^3/_4$, kagomé (ABA'B')	—
Cu$_{13}$(OH)$_{10}$F$_4$[VO$_4$]$_4$ [38]	a 5.802(2) α 110.043(3) b **10.239(4)** β 104.320(4) c 10.914(5) γ 96.662(3)	$P\bar{1}$ 675.6 2	$^3/_4$, triangular stepped stripes separated by honeycombs; triangular net; (AB)	Antiferromagnetic ordering at 3 K

* Plus other isotypic BaMe_2[TO_4]$_2$ (Me = Co, Ni; T = P, V, As) compounds [39–41].

3. The Mero-Plesiotype Series of Structurally-Related Phases

All crystal structures of the series can be represented as an assembly of alternating two-dimensional core module A and other slabs of diverse composition and design (Table 1).

The crystal structure of the "mineralogically probable" divanadate carbonate $K_2Mn_3[VO_4]_2(CO_3)$ [23] is formed by two types of modules alternating along the c axis of the hexagonal unit cell (Figure 4a). Gibbsite-like layer parallel to the ab plane has a honeycomb arrangement of Mn^{2+} cations (Figure 3a) at the centers of octahedra sharing cis and trans edges (Figure 5a). The [VO_4] tetrahedra at both sides of the dioctahedral layer complete the main block to the $Mn_2[VO_4]_2$ composition and provide its linkage along the c axis with the second $\{K_2MnCO_3\}$ module formed by Mn-centered trigonal bipyramids, [CO_3] triangles, and K atoms. The $MnCO_3$ cellular layer is topologically identical in the ab plane to the kalsilite network built of Si and Al tetrahedra (Figure 5b). From both sides, these layers are attached by sheets of K nine-vertex polyhedra sharing vertices. Thus, the structure can be represented as an alteration of negatively charged $\{Mn_2[VO_4]_2\}^{2-}$ and positively charged $\{K_2MnCO_3\}^{2+}$ modules. Recently published, two novel vanadate carbonates $K_2Co_3[VO_4]_2(CO_3)$ and $Rb_2Mn_3[VO_4]_2(CO_3)$ have similar crystal structures [28]. Moreover, the Mn and Co containing formula analogues are isotypic, while the Rb, Mn variety possess trigonal symmetry. The authors of [28] noted the effect of the larger size of Rb^+ in comparison with the size of K^+ on the structure transformation that occurs when the [VO_4] tetrahedra adjacent along the c axis rotate around the [001] direction by 180°. Thus, two vanadate tetrahedra of $\{Me_2[VO_4]_2\}$ modules adjacent along the c axis have the same vertex orientation "up, up" (or "down, down") in the hexagonal $K_2Mn_3[VO_4]_2(CO_3)$ structure, while the opposite "up, down" orientation of these tetrahedra characterizes the trigonal Rb formula analogue (Figure 4). Nevertheless, both compounds have similar to the $K_2Mn_3[VO_4]_2(CO_3)$ structural blocks, alternating $\{Me_2[VO_4]_2\}$ (A) and $\{A_2MeCO_3\}$ (B) modules. In accordance with the action of the 6_3 or $\bar{3}$ axis, the symmetrically multiplied structural A′B′ fragment is shifted by $^1/_2$ of the translation vector along the [001] direction. The sequence of the modules' alternation (ABA′B′) corresponds to the value of the c axis of the unit cell that is in the range from 22 to 23 Å.

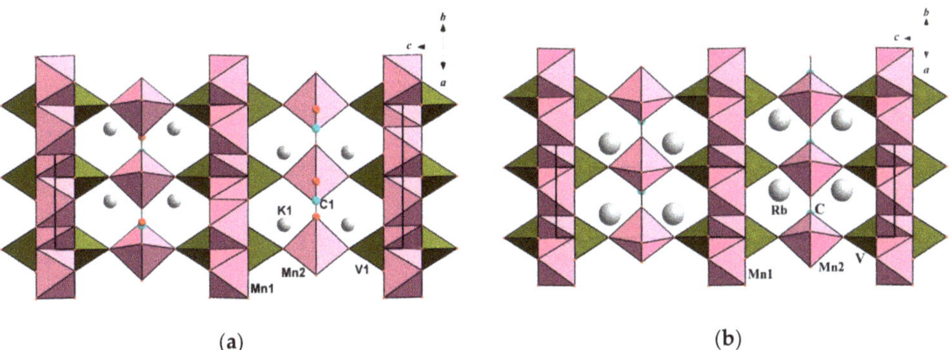

(a) (b)

Figure 4. $K_2Mn_3[VO_4]_2(CO_3)$ (**a**) and $Rb_2Mn_3[VO_4]_2(CO_3)$ (**b**) crystal structures in the [110] projection, showing the different vertex orientation of vanadate tetrahedra.

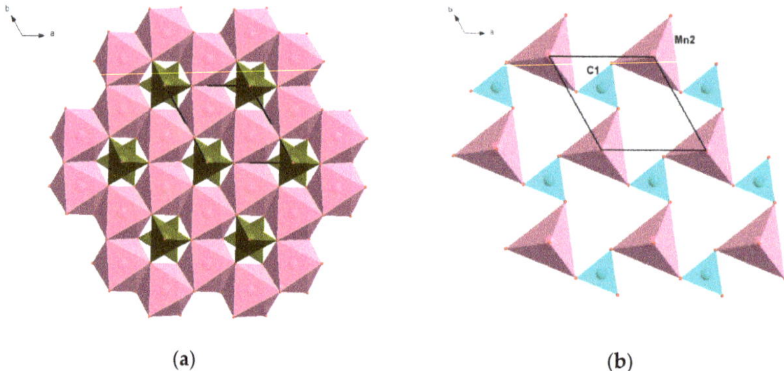

(a)   (b)

Figure 5. Gibbsite-like (**a**) and kalsilite-like (**b**) layers in the *ab* projection of the $K_2Mn_3[VO_4]_2(CO_3)$ crystal structure.

The same blocks built of octahedral layers of the gibbsite-type and adjacent $[VO_4]$ tetrahedra, form the crystal structure of $BaNi_2[VO_4]_2$ [29] with similar parameters of the trigonal unit cell, but with the R Bravais lattice (Table 1). Due to the R lattice translation, neighboring blocks of the $Ni_2[VO_4]_2$ composition are shifted in the $[\bar{1}10]$ direction. Along the c axis, they alternate with layers of large barium 12-vertices polyhedra sharing edges (Figure 6). In the structures of vanadate-carbonates, similar blocks alternate with the $\{A_2MeCO_3\}$ modules, as shown above. The formal transformation of the structure can be restored as a result by an extraction of $[MeCO_3]_\infty$ layers from the $A_2Me_3[VO_4]_2(CO_3)$ crystal structure with the simultaneous exchange of Me^{2+} ions in octahedra for Ni^{2+}, and one Ba^{2+} for two K^+ or two Rb^+ ions. Then, the $BaNi_2[VO_4]_2$ structure is obtained as a derivative of the vanadate-carbonate architecture. Translation period for the module alternation ($ABA'B'A''B''$) defines the unit cell c parameter equal to 22.3 Å (Table 1). The same structural features demonstrate isotypic cobalt phosphate $BaCo_2[PO_4]_2$ [30] and arsenate $BaCo_2[AsO_4]_2$ [31].

Let us mention here two isotypic arsenates, $KNi[AsO_4]$ and $NaNi[AsO_4]$ [32] which crystallize in the same $R\bar{3}$ space group inherent to the $BaNi_2[VO_4]_2$, but their crystal structures include twice more alkali cations between $\{Me_2[TO_4]_2\}$ modules as compared with the amount of alkali earth Ba atoms in the vanadate. Consequently, the increased c parameters of $KNiAsO_4$ and $NaNiAsO_4$ to 28.53 and 26.47 Å correspond to two-layer blocks of sharing edges seven-vertex K- or Na-centered polyhedra between the main modules of the arsenate structures.

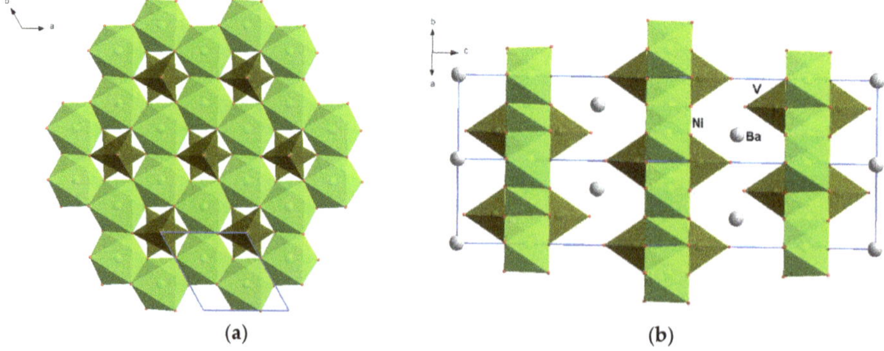

(a)   (b)

Figure 6. The $\{Ni_2[VO_4]_2\}^{2-}$ module based on a gibbsite-type layer in the crystal structure of $BaNi_2[VO_4]_2$ (**a**) and its alternation with Ba-cations in [110] projection (**b**).

Synthetic phases $Na_2Ni_3(OH)_2[PO_4]_2$ [24] and $K_2Mn_3(OH)_2[VO_4]_2$ [36] present an example of isotypic compounds from different chemical classes. A virtual exchange of all cations (except H^+) at four symmetrically independent structural positions (Na^+, $Ni1^{2+}$, $Ni2^{2+}$, and P^{5+}) in the $Na_2Ni_3(OH)_2[PO_4]_2$ structure for cations in the same oxidation state of a larger radius, (K^+, $Mn1^{2+}$, $Mn2^{2+}$, and V^{5+}) leads to the formation of the isostructural phase, $K_2Mn_3(OH)_2[VO_4]_2$, of the same symmetry (space group $C2/m$), but with obviously increased values of the unit cell parameters and the monoclinic angle (Table 1). In both cases, the layers of edge-sharing $MeO_4(OH)_2$ octahedra include $\frac{1}{4}$ of vacancies giving rise to the cationic substructure formed of stripes of the triangular net separated by honeycombs (Figure 3c). Distances between the modules with central layers built of Ni or Mn octahedra are 4.2 and 4.8 Å accordingly, and depend mainly on the sizes of $[PO_4]$ and $[VO_4]$ tetrahedra and the hydrogen bond lengths. The naturally smaller space between neighboring $\{Ni_3(OH)_2[PO_4]_2\}$ blocks in the phosphate structure suits perfectly for being filled by Na atoms (Figure 7). The similar modules alternate with the layers of KO_7 or NaO_7 polyhedra of the same topology (Figures 7 and 8).

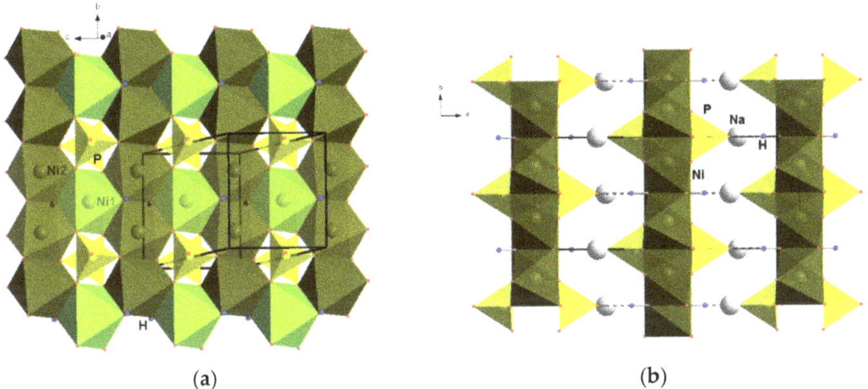

Figure 7. The crystal structure of $Na_2Ni_3(OH)_2[PO_4]_2$ in [101] (**a**) and [001] (**b**) projections.

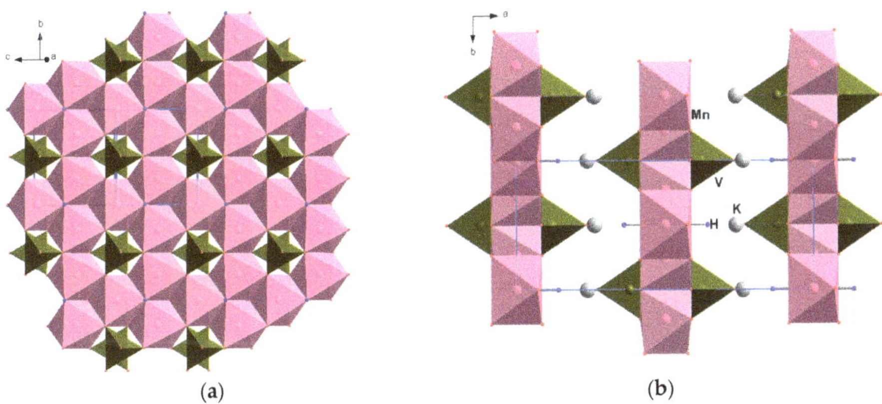

Figure 8. The crystal structure of $K_2Mn_3(OH)_2(VO_4)_2$ in [101] (**a**) and [001] (**b**) projections.

The described above lattice (Figure 3c) stands between the honeycomb lattice (Figure 3a), characteristic of the recently discussed divanadate carbonates, and the kagomé lattice (Figure 3b), inherent in the isostructural synthetic phase $BaNi_3(OH)_2[VO_4]_2$ [34] and mineral vésignéite, $BaCu_3(OH)_2[VO_4]_2$ [33]. In the vésignéite structure, the octahedral voids of the central layers of the main modules are also filled for $\frac{3}{4}$ by the Ni or Cu atoms,

but the distribution of the empty voids inside the octahedral layers corresponds to th
regular kagomé lattice (Figure 9a). Along the [001] direction, these modules (A) alternate
with layers formed by the large Ba-centered polyhedra presenting the second-type slabs
(Figure 9b). The AB sequence of their repetition along the [001] direction defines the valu
of the *c* axis close to 8 Å for both compounds (Table 1).

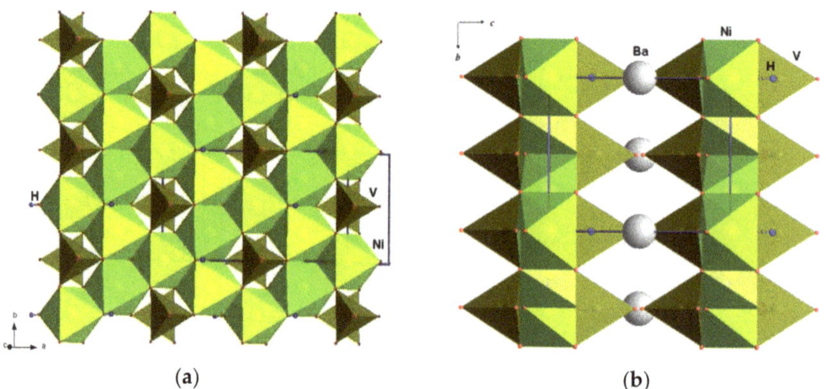

(**a**) (**b**)

Figure 9. The crystal structure of vésignéite, $BaCu_3(OH)_2[VO_4]_2$ in projections onto the *ab* (**a**) and *bc* (**b**) planes.

The cationic arrangement inside the layer centering the basic block of the
$\{(Cu,Zn)_3(OH)_2[AsO_4]_2\}$ composition in the monoclinic crystal structure of mineral bayl
donite, $Pb(Cu,Zn)_3(OH)_2[AsO_4]_2$ [37] also shows the layers with $^3/_4$ of octahedral voids
populated by the Cu/Zn atoms. It corresponds to the same ordinary kagomé configuratior
as in vésignéite (Figure 3b). Along the [001] direction, these blocks (A) alternate with layers
of Pb eight-vertex polyhedra (tetragonal antiprisms) (B) (Figure 10). However, differently
to the similarly monoclinic (sp. gr. $C2/m$) vésignéite structure, in the present case, the
modules are twice multiplied along the *c* axis due to the symmetry (sp. gr. $C2/c$), forming
the ABA'B' sequence with a period of 14 Å.

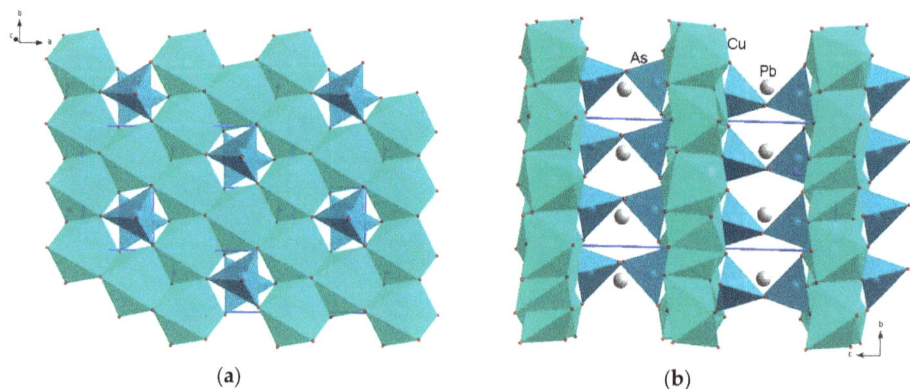

(**a**) (**b**)

Figure 10. The crystal structure of bayldonite, $Pb(Cu,Zn)_3(OH)_2[AsO_4]_2$ projected onto the *ab* (**a**) and *bc* (**b**) planes.

A unique distribution of the filled and empty octahedral voids inside the closest packing
of oxygen atoms is established in the crystal structure of the synthetic oxyfluoride vanadate
$Cu_{13}(OH)_{10}F_4[VO_4]_4$ [38]. Here, the core module A includes $^1/_4$ of the empty octahedra,
but differently to the structures discussed above, an exclusive cationic arrangement within
the layer arises. It can be described as stepped stripes of the triangular net separated by
honeycombs (Figure 3e). The negatively charged A modules $\{Cu_3(OH)_2[VO_4]_2\}^{2-}$ formed

by such octahedral layers with adjacent vanadate tetrahedra, alternate along the *b* axis with B fragments of the brucite-type structure with a triangular sublattice (Figure 3d). These positively charged B-modules $\{Cu_{3.5}(OH)_3F_2\}^{2+}$ are based on the densely packed oxygen and fluorine atoms, with all octahedral voids occupied by Cu. The AB repetition of the modules fixes the value near 10.2 Å of the *b* parameter (Figure 11) of the triclinic unit cell (Table 1).

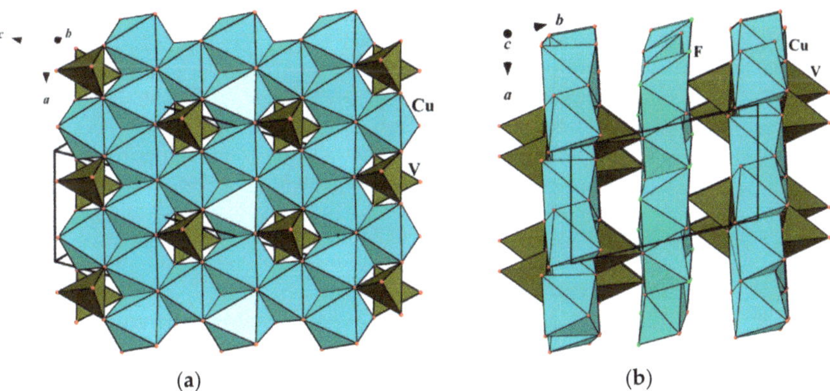

(a)   (b)

Figure 11. The $\{Cu_3(OH)_2[VO_4]_2\}^{2-}$ module in the crystal structure of $Cu_{13}(VO_4)_4(OH)_{10}F_4$ (**a**) and its alternation with octahedral $\{Cu_{3.5}(OH)_3F_2\}^{2+}$ layers in [001] projection (**b**).

4. Magnetic Behavior of the Series Members

Recently, studies of quantum-spin systems have been of considerable interest. On the basis of low-dimensional magnetic systems, exotic quantum ground states of matter can be realized including gapped and gapless spin liquids [42], spin ice [43,44], spin glass, and various peculiar forms of long-range magnetic order [45–47]. If the way of the distribution of spin carriers can be described by a special lattice constructed of triangles sharing vertices or edges, such as kagomé and triangular lattices [48–50], the three-dimensional magnetic order tends to be suppressed due to spin frustration [38]. The presence of ions with an open shell Mn^{2+}, Ni^{2+}, Co^{2+}, or Cu^{2+} inside the octahedral layers causes the appearance of two-dimensional antiferromagnetic or frustrated magnetic properties for the compounds considered in this work. Anionic tetrahedra $[PO_4]^{3-}$, $[VO_4]^{3-}$, or $[AsO_4]^{3-}$ serve as ideal non-magnetic separators between transition-metal 2D planes and contribute to the octahedral environment of oxygen atoms around cations. Moreover, these units can prevent distortions of the *Me*-centered octahedra to preserve the planar geometry of the magnetic substructures. The way in which magnetically active ions fill octahedral voids inside the closely packed oxygen/fluorine atoms determines the type of magnetic structure (Figure 3) that inevitably contains geometric frustrations.

The kagomé lattice should provide a much stronger frustration than the simplest triangular lattice (due to the chirality degeneration), and, as a consequence, the absence of long-range magnetic order [51]. The classical kagomé lattices describe the cationic substructures of the vésignéite $BaCu_3(OH)_2[VO_4]_2$ and $BaNi_3(OH)_2[VO_4]_2$. At 53 K, vésignéite demonstrates a strong antiferromagnetic interaction between spins of nearest-neighbors without long-range magnetic order up to 2 K, and its ground state is assumed to be a gapless spin liquid [52]. In $BaNi_3(OH)_2[VO_4]_2$, magnetic frustration arises due to the competition between ferromagnetic and antiferromagnetic ordering, which leads to a glassy transition at 19 K [34]. According to [51], in the honeycomb-type substructure should be no frustration, except in the presence of next-nearest-neighbor exchange interactions. The honeycomb lattices characterize isostructural $BaMe_2[TO_4]_2$ (*Me* = Co, Ni; *T* = P, V, As) compounds, which display quite different magnetic properties. Thus, $BaNi_2[VO_4]_2$ demonstrates the onset of antiferromagnetic long-range ordering near 50 K [29]. $BaCo_2[PO_4]_2$ is a rare highly frustrated, quasi-2D magnetic material with the honeycomb substructure that displays competing for short-range magnetic ordering below $T_{N1} \sim 6$ K and $T_{N2} \sim 3.5$ K, but resists long-range magnetic order and spin freezing [44]. The complex magnetic ordering and spin dynamics in $BaCo_2[AsO_4]_2$ has been investigated several times, but still remains an open problem. The $BaCo_2[AsO_4]_2$ frustrated magnet with the honeycomb distribution of Co atoms sharply ordered at T ~ 5.4 K with a probable formation of a helical design [53], but repeated research revealed its quasi-collinear incommensurate ground-state structure [54]

Another type of frustrated substructure, described as built from stripes of triangular nets separated by honeycombs, which is a compromise between honeycomb and kagomé lattices, should not have an antiferromagnetic order of the Néel-type, but should have geometric magnetic frustration [51]. The same lattice characterizes the isotypic crystal structures of $Na_2Ni_3(OH)_2(PO_4)_2$, $Na_2Co_3(OH)_2[VO_4]_2$, and $K_2Mn_3(OH)_2[VO_4]_2$; they all demonstrate low-dimensional antiferromagnetic ordering of various origins. In the phosphate compound, it arises as a result of the presence and competition of ferro- and antiferromagnetic interactions [24]. Mn vanadate exhibits geometric frustration and low-dimensionality effects and spin–lattice coupling [51]. Cobalt phase shows evidence of spin–orbit coupling in Co^{2+} ions with antiferromagnetic ordering at 4.4.K and highly anisotropic field-dependent behavior with multiple metamagnetic transitions [35]. Dual-width stripes of triangular net are surrounded by honeycombs in the substructure of reppiaite, $Mn_5(OH)_4[VO_4]_2$. This 2D magnet displays Curie−Weiss behavior above 100 K with significant antiferromagnetic coupling and canted antiferromagnetic order below 57 K [55].

In crystal structures of some compounds, magnetic cations form two types of magnetic arrangements. Thus, the complex structure of $Cu_{13}(OH)_{10}F_4[VO_4]_4$ is described by the triangular lattice, which alternates with a unique substructure of dual-width stripes of the triangular net separated by honeycombs. The magnetic system of $Cu_{13}(OH)_{10}F_4[VO_4]_4$ created in this way exhibits long-range antiferromagnetic ordering at $\sim$3 K, a strong spin-frustration effect, and a spin-flop transition at 5 T [38]. Each member of the vanadate-carbonate family is also characterized by two magnetic subsystems, honeycomb- and triangular-type in their crystal structures. They all order antiferromagnetically but at different temperatures due to diverse sorts of transition cations. $K_2Mn_3[VO_4]_2(CO_3)$ showed a two-step formation of long-range magnetic order at low temperatures [23]. The following study showed that triangular and honeycomb magnetic layers undergo sequential magnetic ordering and act as nearly independent magnetic subsystems. The honeycomb substructure orders at about 85 K in a Néel-type antiferromagnetic structure, while the triangular arrangement displays two consecutive ordered states at much lower temperatures of 3 and 2.2 K [56]. Likewise, the Rb-analogue $Rb_2Mn_3[VO_4]_2(CO_3)$ exhibits three magnetic transitions at 77 K, 2.3 K, and 1.5 K. At 77 K, it orders in the honeycomb layer in a Néel-type antiferromagnetic orientation, while the lower temperature spin structure has either a collinear or a canted magnetic structure for the triangular Mn lattice [28]. $K_2Co_3(VO_4)_2CO_3$ displays a canted antiferromagnetic ordering below $T_N = 8$ K [28].

5. Conclusions

We have shown [57] that the crystal structures of the mero-plesiotype series of natural and synthetic vanadates, phosphates and arsenates with first row transition metals are formed of similar 2D modules, which alternate in one direction with other structural fragments, diverse for different members of the series. The core modules are built of a central octahedral layer filled with *d* elements (Mn, Ni, Cu, or Co) and adjacent anionic (VO_4, PO_4, AsO_4) tetrahedra. The central layer is based on the closest packing of oxygen (and fluorine) atoms. Changed amounts of the octahedral voids and their varying distribution differ from one structure to another in a framework of the same anionic substructure. Seven "magnetic" topologies of the transition metal distribution within the layers have been identified: triangular, honeycomb, kagomé, and different combinations of fragments of the triangular and honeycomb lattices (Figure 3). The crystal structure constitution based on the same repeating modules establishes the 2D character design for all polysomes. Consequently, they all have similar translations inside the module plane equal of about 5 Å, while the values of the unit cell axis perpendicular to the sheet is defined by the size of the second module and the sequence of the module's alternation. The only basic type of modules with similar dimensions reproduces in one direction in the structures of minerals reppiaite $Mn_5[VO_4]_2(OH)_4$ and cornubite $Cu_5(OH)_4[AsO_4]_2$, as well as synthetic arsenate $Ni_5(OH)_4[AsO_4]_2$. Therefore, these phases with the crystal structures built exclusively from the main modules, on which the polysomatic series of vanadates, arsenates, and phosphates is based, represent the archetype structures.

When additional (B) slabs of the alkaline (or alkaline-earth) metals interrupt the basic (A) modules, a few combinations within the mero-plesiotype series are formed: (AB) for vésignéite $BaCu_3(OH)_2[VO_4]_2$ and its Ni-analogue $BaNi_3(OH)_2[VO_4]_2$ with approximately 7.8 Å period along the modules' alternation, (ABA'B') for bayldonite $Pb(Cu,Zn)_3(OH)_2[AsO_4]_2$, $Na_2Ni_3(OH)_2(PO_4)_2$ and $K_2Mn_3(OH)_2[VO_4]_2$ with about 14.5 Å period, and (ABA'B'A''B'') for $BaNi_2[VO_4]_2$ with 22.3 Å translation. Another kind of supplementary slabs present the octahedral layers of brucite topology, which interconnect with the basic modules in the $Cu_{13}(OH)_{10}F_4[VO_4]_4$ crystal structure resulting the (AB) translation period equal to 10.2 Å. In the crystal structures of vanadate-carbonates $K_2Mn_3[VO_4]_2(CO_3)$, $K_2Co_3[VO_4]_2(CO_3)$, and $Rb_2Mn_3[VO_4]_2(CO_3)$, two different modules alternate in a sequence (ABA'B') along the c axis of about 22.4 Å. The main block is a core module based on the gibbsite-type octahedral layer. The second one is the block formed by the kalsilite-like sheet of Mn bipyramids and CO_3 triangles that is sandwiched between two layers of alkaline metals.

It is important to note that the minerals and laboratory-synthesized compounds considered here exhibit magnetic properties, representing two-dimensional antiferromagnets or frustrated magnets. Obviously, their magnetic behavior is directly related to the crystal structure peculiarities. In particular, the way in which magnetically active Mn^{2+}, Ni^{2+}, Co^{2+}, or Cu^{2+} ions occupy the octahedral voids between the densely packed oxygen atoms is crucial.

Author Contributions: Conceptualization, O.Y.; validation, O.Y.; investigation, O.Y. and G.K.; writing—original draft preparation, G.K.; writing—review and editing, O.Y. All authors have read and agreed to the published version of the manuscript.

Funding: The work was carried out with the support of a grant from the President of the Russian Federation for young scientists-candidates of science MK−1613.2021.1.5.

Informed Consent Statement: Not applicable.

Data Availability Statement: Not applicable.

Acknowledgments: We are much obliged to L.V. Shvanskaya for valuable recommendations and comments. We thank G. Ferraris and the reviewers for their constructive criticism, which definitely improved the article.

Conflicts of Interest: The authors declare no conflict of interest.

References

1. Thompson, J.B., Jr. Geometrical possibilities for amphibole structures: Model biopyriboles. *Amer. Miner.* **1970**, *55*, 292–293.
2. Thompson, J.B. An Introduction to the Mineralogy and Petrology of the Biopyriboles. In *Amphiboles and Other Hydrous Pyriboles—Mineralogy, Reviews in Mineralogy*; Veblen, D.R., Ed.; Mineralogical Society of America: Washington, DC, USA, 1981; Volume 9A, pp. 141–188.
3. Thompson, J.B., Jr. Biopyriboles and polysomatic series. *Amer. Miner.* **1978**, *63*, 239–249.
4. Veblen, D.R. Non-Classical Pyriboles and Polysomatic Reactions in Biopyriboles. In *Amphiboles and other Hydrous Pyriboles—Mineralogy, Reviews in Mineralogy*; Veblen, D.R., Ed.; Mineralogical Society of America: Washington, DC, USA, 1981; Volume 9A, pp. 189–236.
5. Makovicky, E. Modularity—Different Types and Approaches. In *EMU Notes in Mineralogy, Modular Aspects of Minerals*; Merlino, S., Ed.; Eötvös University Press: Budapest, Hungary, 1997; Volume 1, pp. 315–343.
6. Ferraris, G. Polysomatic Aspects of Microporous Minerals—Heterophyllosilicates, Palysepioles and Rhodesite-Related Structures. *Rev. Miner. Geochem.* **2005**, *57*, 69–104. [CrossRef]

7. Cadoni, M.; Ferraris, G. Two new members of the rhodesite mero-plesiotype series close to delhayelite and hydrodelhayelite. Synthesis and crystal structure. *Eur. J. Miner.* **2009**, *21*, 485–493. [CrossRef]
8. Veblen, D.R. Polysomatism and polysomatic series: A review and application. *Amer. Miner.* **1991**, *76*, 801–826.
9. Veblen, D.R.; Buseck, P.R. Microstructures and reaction mechanisms in biopyriboles. *Amer. Miner.* **1980**, *65*, 599–623.
10. Welch, M.; Klinowski, J. Characterization of polysomatism in biopyriboles: Double-/triple-chain lamellar intergrowths. *Phy. Chem. Miner.* **1992**, *18*, 460–468. [CrossRef]
11. Merlino, S.; Pasero, M. Polysomatic Approach in the Crystal Chemistry Study of Minerals. In *EMU Notes in Mineralogy, Modular Aspects of Minerals*; Merlino, S., Ed.; Eötvös University Press: Budapest, Hungary, 1997; Volume 1, pp. 297–312.
12. Baronnet, A.; Papp, G.; Merlino, S. Equilibrium and Kinetic Processes for Polytype and Polysome Generation. In *Modular Aspects of Minerals*; Mineralogical Society of America: Chantilly, VA, USA, 1997; Volume 1, pp. 119–152.
13. Drits, V.A.; Papp, G.; Merlino, S. Mixed-layer Minerals. In *Modular Aspects of Minerals*; Mineralogical Society of America: Chantilly, VA, USA, 1997; Volume 1, pp. 153–190.
14. Ferraris, G.; Makovicky, E.; Merlino, S. *Crystallography of Modular Materials*; Oxford University Press (OUP): Oxford, UK, 2008.
15. Zvyagin, B.B.; Papp, G.; Merlino, S. Modular Analysis of Crystal Structures. In *Modular Aspects of Minerals*; Mineralogical Society of America: Chantilly, VA, USA, 1997; Volume 1, pp. 345–372.
16. Ferraris, G.; Ivaldi, G. Structural Features of Micas. *Rev. Miner. Geochem.* **2002**, *46*, 117–153. [CrossRef]
17. Nespolo, M.; Ferraris, G.; Ďurovič, S.; Takeuchi, Y. Twins vs. modular crystal structures. *Z.Kristallogr. Cryst. Mater.* **2004**, *219*, 773–778. [CrossRef]
18. Ferraris, G. Modular structures The paradigmatic case of the heterophyllosilicates. *Z. Kristallogr. Cryst. Mater.* **2008**, *223*, 76–84. [CrossRef]
19. Yakubovich, O.V.; Dem'yanetz, L.N.; Massa, W. A New Cu, Al Fluoride Disilicate $CuAl_2F_2(Si_2O_7)$ and its Relations to Topaz. *Z. Anorg. Allg. Chem.* **2000**, *626*, 1514–1518. [CrossRef]
20. Massa, W.; Yakubovich, O.V.; Kireev, V.V.; Mel'Nikov, O.K. Crystal structure of a new vanadate variety in the lomonosovite group $Na_5Ti_2O_2[Si_2O_7](VO_4)$. *Solid State Sci.* **2000**, *2*, 615–623. [CrossRef]
21. Yakubovich, O.V.; Massa, W.; Chukanov, N.V. Crystal structure of britvinite $[Pb_7(OH)_3F(BO_3)_2(CO_3)][Mg_{4.5}(OH)_3(Si_5O_{14})]$: A new layered silicate with an original type of silicon-oxygen networks. *Crystallogr. Rep.* **2008**, *53*, 206–215. [CrossRef]
22. Bozhilov, K.N. Structures and Microstructures of Non-Classical Pyriboles. In *EMU Notes in Mineralogy, Minerals at the Nanoscale*; Nieto, F., Livi, K.J.T., Oberti, R., Eds.; European Mineralogical Union and the Mineralogical Society of Great Britain & Ireland: London, UK, 2013; Volume 14, pp. 109–152.
23. Yakubovich, O.V.; Yakovleva, E.V.; Golovanov, A.N.; Volkov, A.S.; Volkova, O.S.; Zvereva, E.A.; Dimitrova, O.V.; Vasiliev, A.N. The First Vanadate–Carbonate, $K_2Mn_3(VO_4)_2(CO_3)$: Crystal Structure and Physical Properties. *Inorg. Chem.* **2013**, *52*, 1538–1543. [CrossRef]
24. Yakubovich, O.; Kiriukhina, G.; Dimitrova, O.; Volkov, A.; Golovanov, A.; Volkova, O.; Zvereva, E.; Baidya, S.; Saha-Dasgupta, T.; Vasiliev, A. Crystal structure and magnetic properties of a new layered sodium nickel hydroxide phosphate, $Na_2Ni_3(OH)_2(PO_4)_2$. *Dalton Trans.* **2013**, *42*, 14718–14725. [CrossRef] [PubMed]
25. Basso, R.; Lucchetti, G.; Zefiro, L.; Palenzona, A. Reppiaite, $Mn_5(OH)_4(VO_4)_2$, a new mineral from Val Graveglia (Northern Apennines, Italy). *Z. Kristallogr. Cryst. Mater.* **1992**, *201*, 223–234. [CrossRef]
26. Tillmanns, E.; Hofmeister, W.; Petitjean, K. Cornubite, $Cu_5(AsO_4)_2(OH)_4$, first occurrence of single crystals, mineralogical description and crystal structure. *Bull. Geol. Soc. Finl.* **1985**, *57*, 119–127. [CrossRef]
27. Barbier, J. The crystal structure of $Ni_5(AsO_4)_2(OH)_4$ and its comparison to other $M_5(XO_4)_2(OH)_4$ compounds. *Eur. J. Miner.* **1996**, *8*, 77–84. [CrossRef]
28. Pellizzeri, T.M.S.; Sanjeewa, L.D.; Pellizzeri, S.; McMillen, C.D.; Garlea, V.O.; Ye, F.; Sefat, A.S.; Kolis, J.W. Single crystal neutron and magnetic measurements of $Rb_2Mn_3(VO_4)_2CO_3$ and $K_2Co_3(VO_4)_2CO_3$ with mixed honeycomb and triangular magnetic lattices. *Dalton Trans.* **2020**, *49*, 4323–4335. [CrossRef]
29. Rogado, N.; Huang, Q.; Lynn, J.W.; Ramirez, A.P.; Huse, D.; Cava, R.J. BaNi2V2O8: A two-dimensional honeycomb antiferromagnet. *Phys. Rev. B* **2002**, *65*, 144443. [CrossRef]
30. Bircsak, Z.; Harrison, W.T.A. Barium Cobalt Phosphate, $BaCo_2(PO_4)_2$. *Acta Crystallogr. Sect. C Cryst. Struct. Commun.* **1998**, *54*, 1554–1556. [CrossRef]
31. Đordević, T. $BaCo_2(AsO_4)_2$. *Acta Crystallogr. Sect. E Struct. Rep. Online* **2008**, *64*, i58–i59. [CrossRef]
32. Buckley, A.M.; Bramwell, S.T.; Day, P.; Harrison, W.T.A. The Crystal Structure of Potassium Nickel Arsenate; $KNiAsO_4$. *Z. Naturforsch. B* **1988**, *43*, 1053–1055. [CrossRef]
33. Zhesheng, M.; Ruilin, H.; Xiaoling, Z. Redetermination of the Crystal Structure of Vesignieite. *Acta Geol. Sin.* **2009**, *4*, 145–151. [CrossRef]
34. Freedman, D.E.; Chisnell, R.; McQueen, T.M.; Lee, Y.S.; Payen, C.; Nocera, D.G. Frustrated magnetism in the S = 1 kagomé lattice $BaNi_3(OH)_2(VO_4)_2$. *Chem. Commun.* **2011**, *48*, 64–66. [CrossRef]
35. Pellizzeri, T.M.S.; Morrison, G.; McMillen, C.D.; Loye, H.Z.; Kolis, J.W. Sodium Transition Metal Vanadates from Hydrothermal Brines: Synthesis and Characterization of $NaMn_4(VO_4)_3$, $Na_2Mn_3(VO_4)_3$, and $Na_2Co_3(VO_4)_2(OH)_2$. *Eur. J. Inorg. Chem.* **2020**, *2020*, 3408–3415. [CrossRef]

36. Liao, J.-H.; Guyomard, D.; Piffard, Y.; Tournoux, M. $K_2Mn_3(OH)_2(VO_4)_2$, a New Two-Dimensional Potassium Manganese(II) Hydroxyvanadate. *Acta Crystallogr. Sect. C Cryst. Struct. Commun.* **1996**, *52*, 284–286. [CrossRef]
37. Ghose, S.; Wan, C. Structural chemistry of copper and zinc minerals. VI. Bayldonite, $(Cu,Zn)_3Pb(AsO_4)_2(OH)_2$: A complex layer structure. *Acta Crystallogr. Sect. B Struct. Crystallogr. Cryst. Chem.* **1979**, *35*, 819–823. [CrossRef]
38. Yang, M.; Zhang, S.-Y.; Guo, W.-B.; Tang, Y.-Y.; He, Z.-Z. Spin-frustration in a new spin-1/2 oxyfluoride system $(Cu_{13}(VO_4)_4(OH)_{10}F_4)$ constructed by alternatively distorted kagome-like and triangular lattices. *Dalton Trans.* **2015**, *44*, 15396–15399. [CrossRef]
39. El-Bali, B.; Bolte, M.; Boukhari, A.; Aride, J.; Taibe, M. $BaNi_2(PO_4)_2$. *Acta Crystallogr. Sect. C Cryst. Struct. Commun.* **1999**, *55*, 701–702. [CrossRef]
40. Régnault, L.; Henry, J.; Rossat-Mignod, J.; De Combarieu, A. Magnetic properties of the layered nickel compounds $BaNi_2(PO_4)_2$ and $BaNi_2(AsO_4)_2$. *J. Magn. Magn. Mater.* **1980**, *15–18*, 1021–1022. [CrossRef]
41. Zhong, R.; Chung, M.; Kong, T.; Nguyen, L.T.; Lei, S.; Cava, R.J. Field-induced spin-liquid-like state in a magnetic honeycomb lattice. *Phys. Rev. B* **2018**, *98*, 220407. [CrossRef]
42. Balents, L. Spin liquids in frustrated magnets. *Nat. Cell Biol.* **2010**, *464*, 199–208. [CrossRef] [PubMed]
43. Bramwell, S.T. Spin Ice State in Frustrated Magnetic Pyrochlore Materials. *Science* **2001**, *294*, 1495–1501. [CrossRef] [PubMed]
44. Nair, H.S.; Brown, J.M.; Coldren, E.; Hester, G.; Gelfand, M.P.; Podlesnyak, A.; Huang, Q.; Ross, K.A. Short-range order in the quantum XXZ honeycomb lattice material $BaCo_2(PO_4)_2$. *Phys. Rev. B* **2018**, *97*, 134409. [CrossRef]
45. Greedan, J.E. Geometrically frustrated magnetic materials. *J. Mater. Chem.* **2001**, *11*, 37–53. [CrossRef]
46. Moessner, R.; Ramirez, A.P. Geometrical frustration. *Phys. Today* **2006**, *59*, 24–29. [CrossRef]
47. Suzuki, N.; Matsubara, F.; Fujiki, S.; Shirakura, T. Absence of classical long-range order in an S=1/2 Heisenberg antiferromagnet on a triangular lattice. *Phys. Rev. B Condens. Matter* **2014**, *90*, 184414. [CrossRef]
48. Collins, M.F.; Petrenko, O.A. Review/Synthèse: Triangular antiferromagnets. *Can. J. Phys.* **1997**, *75*, 605–655. [CrossRef]
49. Harrison, A. First catch your hare: The design and synthesis of frustrated magnets. *J. Physics: Condens. Matter* **2004**, *16*, S553–S572. [CrossRef]
50. Pati, S.K.; Rao, C.N.R. Role of the spin magnitude of the magnetic ion in determining the frustration and low-temperature properties of kagome lattices. *J. Chem. Phys.* **2005**, *123*, 234703. [CrossRef] [PubMed]
51. Otsuka, D.; Sato, H.; Matsuo, A.; Kindo, K.; Nakamura, D.; Takeyama, S. Ultrahigh-Magnetic-Field Magnetization of Multi-Kagome-Strip (MKS) Lattice Spin-Frustrated Magnet $K_2Mn_3(OH)_2(VO_4)_2$. *J. Phys. Soc. Jpn.* **2018**, *87*, 124701. [CrossRef]
52. Okamoto, Y.; Yoshida, H.; Hiroi, Z. Vesignieite $BaCu_3V_2O_8(OH)_2$ as a Candidate Spin-1/2 Kagome Antiferromagnet. *J. Phys. Soc. Jpn.* **2009**, *78*, 033701. [CrossRef]
53. Regnault, L.; Burlet, P.; Rossat-Mignod, J. Magnetic ordering in a planar X—Y model: $BaCo_2(AsO_4)_2$. *Phys. B+C* **1977**, *86–88*, 660–662. [CrossRef]
54. Regnault, L.-P.; Boullier, C.; Lorenzo, J. Polarized-neutron investigation of magnetic ordering and spin dynamics in $BaCo_2(AsO_4)_2$ frustrated honeycomb-lattice magnet. *Heliyon* **2018**, *4*, e00507. [CrossRef] [PubMed]
55. Sanjeewa, L.D.; McGuire, M.A.; McMillen, C.D.; Willett, D.; Chumanov, G.; Kolis, J.W. Honeycomb-like S = 5/2 Spin–Lattices in Manganese(II) Vanadates. *Inorg. Chem.* **2016**, *55*, 9240–9249. [CrossRef] [PubMed]
56. Garlea, V.O.; Sanjeewa, L.D.; McGuire, M.A.; Batista, C.D.; Samarakoon, A.M.; Graf, D.; Winn, B.; Ye, F.; Hoffmann, C.; Kolis, J.W. Exotic Magnetic Field-Induced Spin-Superstructures in a Mixed Honeycomb-Triangular Lattice System. *Phys. Rev. X* **2019**, *9*, 011038. [CrossRef]
57. Yakubovich, O.V.; Yakovleva, E.V.; Kirjukhina, G.V. A polysomatic series of two-dimensional vanadates, arsenates and phosphates. *Acta Crystallogr. Sect. A Found. Crystallogr.* **2013**, *69*, 125. [CrossRef]

MDPI
St. Alban-Anlage 66
4052 Basel
Switzerland
Tel. +41 61 683 77 34
Fax +41 61 302 89 18
www.mdpi.com

Minerals Editorial Office
E-mail: minerals@mdpi.com
www.mdpi.com/journal/minerals